T0290961

AGRICULTURE AND ENVIRONMENTAL SECURITY IN SOUTHERN ONTARIO'S WATERSHEDS

AGRICULTURE ISSUES AND POLICIES

Additional books in this series can be found on Nova's website
under the Series tab.

Additional E-books in this series can be found on Nova's website
under the E-book tab.

AGRICULTURE ISSUES AND POLICIES

AGRICULTURE AND ENVIRONMENTAL SECURITY IN SOUTHERN ONTARIO'S WATERSHEDS

GLEN FILSON
EDITOR

Nova Science Publishers, Inc.
New York

For permission to use material from this book please contact us:
Telephone 631-231-7269; Fax 631-231-8175
Web Site: http://www.novapublishers.com

LIBRARY OF CONGRESS CATALOGING-IN-PUBLICATION DATA
Filson, Glen C., 1947-
Agriculture and environmental security in southern Ontario's watersheds / authors, Glen Filson, Bamidele Adekunle, Katia Marzall.
p. cm.
Includes index.
ISBN 978-1-61668-156-2 (hardcover)
1. Agricultural pollution--Ontario, Southern. 2. Water--Pollution--Ontario, Southern. 3. Agriculture--Environmental aspects--Ontario, Southern. I. Adekunle, Bamidele. II. Marzall, Katia. III. Title.
TD195.A34F55 2010
363.739'409713--dc22
2010001734

Published by Nova Science Publishers, Inc. ✛ *New York*

CONTENTS

PREFACE

This book attempts to undertake critical, social scientific scrutiny of contradictions between the evolving forms of agricultural production and ecosystem health within Ontario's southern agricultural watersheds in the light of prevailing environmental management systems and policies.

ACRONYMS

AAFC-Agriculture and Agri-Food Canada
AGCare-Agricultural Groups Concerned About Resources and the Environment
AIC- Agricultural Institute of Canada
APF- Agricultural Policy Framework
ACWF-American's Clean Water Foundation
ABCA-Ausable Bayfield conservation Authority
BMP- Best Management Practices
BNA-British North American Act
EPA-Environmental Protection Act
COFSP-Canada-Ontario Farm Stewardship Program
COWSEP-Canada-Ontario Water Supply Expansion Program
CUSTA-Canada U.S. Free Trade Agreement
CC-Canagagigue Creek
CEECs-Central Eastern European Countries
CFFO-Christian Farmers' Federation of Ontario
CURB-Clean up Rural Beaches Program
CWA-Clean Water Act
CAP-Common Agricultural Policy
CSA-Community Shared Agriculture
DDT- Dichloro-diphenyl-trichloroethane
EAA-Environmental Assessment Act
EBR-Environmental Bill of Rights
EFP-Environmental Farm Plan
EGS-Environmental Goods and Services
EMS-Environmental Management Systems
ESI-Environmental Sustainability Initiative

EQIP-Environmental Quality Incentives Program
EVP-Environmental Voluntary Program
EAEP-Equivalent Agri-Environmental Plan
E/S-Eramosa/Speed Rivers
EIF-European Initiative for Integrating Farming
EU-European Union
FCA-Full cost accounting
GM-Genetically Modified
GRCA-Grand River Conservation Authority
GLWQ-Great Lakes Water Quality Program
GTA-Greater Toronto Area
GC-Greencover Canada
GHG-Greenhouse gases
HACCP-Hazard Analysis and Critical Control Point
EIF-European Initiative for Integrated Farming
IFM-Integrated Farm Management
IPM-Integrated Pest Management
IISD-International Institute for Sustainable Development
ISO-International Organization for Standardization
ILO-Intensive livestock operations
LMAP-Land Management Assistance Program
LS-Land Stewardship Program
LEAF-Linking Environment and Farming
MVCA-Maitland Valley Conservation Authority
MAH-Ministry of Municipal Affairs and Housing
MNR-Ministry of Natural Resources
NAHARP-National Agri-Environmental Health Analysis and Reporting Program
NFSP-National Farm Stewardship Program
NSCP-National Soil Conservation Program
NRCS-Natural Resources Conservation Service
NPS-Non-point source
NAFTA-North American Free Trade Agreement
NMA-Nutrient Management Act
NMP-Nutrient Management Plan
OFAER-On Farm Assessment & Environmental Review Project
OFEC-Ontario Farm Environmental Coalition
OFA-Ontario Federation of Agriculture
OLA-Ontario Landowner's Association

OSCIA-Ontario Soil and Crop Improvement Association
OMAF-Ministry of Agriculture and Food
OMAFRA-Ontario Ministry of Agriculture, Food and Rural Affairs
OFEC-Ontario Farm Coalition
OFAC-Ontario Farm Animal Council
OEFA-Our Farm Environmental Agenda
OSCIA-Ontario Soil and Crop Improvement Association
PA-Planning Act
PP-Precautionary principle
RRCA-Raisin Region Conservation Authority
RWQP-Rural Water Quality Program
RWQ-Rural Water Quality
SFP-Small Farm Plan
SWEEP-Soil and Water Environmental Enhancement Program
SPP-Source Protection Planning
UPA-Union de Producteurs Agricole
UK-United Kingdom
US-United States of America
USDA-United States Development Agency
USEPA-United States Environmental Protection Agency
WAP-Watershed Agricultural Program
WEEP-Whole Farm Easement Program
WFP-Whole Farm Plan
WASI-Wisconsin Agricultural Stewardship Initiative
WTO-World Trade Organization

In: Agricultural and Environmental Security...   ISBN: 978-1-61668-156-2
Editor: Glen Filson   ©2011 Nova Science Publishers, Inc.

Chapter 1

FOOD REGIMES, SUSTAINABILITY AND ONTARIO AGRICULTURE: BOOK OVERVIEW

Glen Filson

INTRODUCTION[1]

With growing demand for food from increasingly prosperous Asian countries at a time of high oil prices and increasing production of ethanol as a substitute, food prices peaked in 2008 accompanied by food riots in many countries (Blas, 2008). As the globalized recession deepened, food prices declined somewhat but the worldwide agri-food crisis continues .

Beside the functions of producing essential food and fiber, there are positive and negative environmental impacts of agriculture. On the one hand, agriculture produces many environmental services by sequestering carbon, preserving biodiversity and helping to manage watersheds. On the other hand, conventional agriculture contributes to climate change via greenhouse gas production; it generates water scarcity, agro-chemical pollution and soil compaction (World Bank, 2007). The rise of intensive agriculture in Ontario has also been accompanied by many environmental problems (Filson, 2004d).

[1] Thanks to the Ontario Ministry of Agriculture, Food and Rural Affairs (OMAFRA) for funding most of the research described in this book.. The Ontario Ministry of the Environment (MOE) funding has provided valuable funding for some of this research as has the Ausable Bayfield Conservation Authority. Thanks too to local Ontario Federation of Agriculture (OFA) representatives who have also assisted with farmer liaison and focus group support. This does not imply endorsement of this work by any of the above.

This book attempts to undertake critical, social scientific scrutiny of contradictions between the evolving forms of agricultural production and ecosystem health within Ontario's southern agricultural watersheds in the light of prevailing environmental management systems and policies.

THE INTERNATIONAL CONTEXT

The global spread of the structural crisis of the capitalist exchange and production system "which is now actively engaged in producing even a global food crisis, on top of all of its other crying contradictions, including the ever more pervasive destruction of nature" cannot be resolved without attempting to address its most profound contradictions (Mészáros, 2008:   10). McMichael (2009:   136) has since observed that "Contradictory relations within food regimes produce crisis, transformation, and transition to successor regimes." So even though the heart of this book deals with what farmers are doing to manage the security of their environments, their local systems cannot be understood apart from the contradictions impacting the latest international food regime or food system within which they must operate.

Friedmann and McMichael (1989) have argued that the first international food regime became evident during the colonial, mercantilist period as produce from the colonies became integrated into European economies. Britain, as the 'workshop of the world' at that time depended on its white settler economies to provide its workers with industrial and wage foods. Since the onset of agricultural industrialization in the early 20th century which saw an influx of technologies into Canadian farming that lowered costs and replaced labour with capital, the rising productivity of farming has helped to integrate Canadian agriculture into the prevailing international food regime (Winson, 1992; Friedmann, and McMichael, 1989).

In the process, small, independent, mixed farming has been increasingly replaced by more commercialized farming producing food and fiber for the world market. At the start of the past century, the introduction of tractors was followed by expanded usage of chemical fertilizers, herbicides and pesticides after WWII, then by food manufacturing in the 1960s and 70s and an accelerated industrialization of agriculture since then.

A new food regime appeared after World War II along with the rise of the American empire when international food regimes became dominated by an aid-based food regime controlled by agro-industrial complexes. Meanwhile, particularly in southwestern Ontario, intensification of production was most notable within intensive livestock production, and this intensification involved an increased concentration and specialization of production (Bowler, 1992; Filson, 2004a).

Well into the 1970s following the rise of what had then been the latest international food regime, the slow decline of the peasantry and continued resilience of small family farms, initially baffled many social scientists (Friedmann, 1978; Vergopoulos, 1978; and Mann and Dickenson, 1978). With the election of Margaret Thatcher in Britain in 1979 and Ronald Reagan in the U. S. in 1981, an era of capitalist deregulation and free trade pitted the world's more corporatized, vertically integrated food producers against less efficient peasants and simple commodity producing 'small' farmers, hastening the latter's decline. As an international farm crisis in the 1980s coincided with increased mobility of money capital, the peasantry and simple commodity producing family farm were increasingly replaced by larger, more commercial farming especially in the United States (Buttel, 2001) and somewhat belatedly in Canada.

Mazoyer and Roudart (2006) argue that the global agrarian crisis was the result of the second agricultural revolution that lowered prices because of rapid productivity gains in the Western world which transformed the gross productivity of the most mechanized and motorized farmers. The result of this increasingly capital intensive farming is that the world's most productive farming was transformed from being ten times greater than that of the least productive at the turn of the 20th century to a factor of one hundred times more productive a century later.

Despite the many natural advantages of the simple, mixed commodity producing family farm (Friedmann, 1986) this industrialization and intensification has steadily lowered food prices and, in turn, ruined producers operating with less mechanized, capitalist production systems (Mazoyer and Roudart, 2006). McMichael (2000) argues that we are now part of an even newer food regime which has arisen along with the 'globalization project' which he argues has largely displaced the post World War II 'development project.'

The world's dominant capitalist powers have worked through the General Agreement on Tariffs and Trade which evolved into the World Trade Organization and a new, deregulated globally mobile international finance system whereby countries produce and export the food that they happen to have a comparative advantage in producing. This has happened using floating exchange rates which were established following the collapse of the earlier

Bretton Woods Accord (1944) of fixed exchange rates when the Americans announced in 1971 that they would end the convertibility of the dollar to gold and many other nations had to float their currencies (Braithwaite and Drahos, 2008). Attempts are now being made at the end of the first decade of the 21st Century, to reinvent a more closely regulated international financial system to avoid a prolonged depression as confidence in the U.S. dollar as a reserve currency declines.

The increasingly globalized international food regime has put added pressure on Ontario farmers to "get big or get out". Surviving small farm operations often have inadequate productivity and annual revenues so they are caught between low margins and rising urban expectations for a more environmentally friendly agriculture (see chapters 3 – 6). Canadian farmers produce reasonably safe and reliable food and fiber while helping to reduce the world's depleted food stocks, yet in Ontario, where the agricultural industry is the second largest provincial industry, the debate about how sustainable its methods are, continues to rage. Even though the industrialization of farming has increased productivity phenomenally, especially among the most industrialized farming operations, the changes that have ensued have also created environmental and social problems as the number of Ontario farms continues to decrease and the remaining farmers must operate more intensively than ever on a declining land base threatened by urban sprawl.

A very substantial part of the rural to urban land conversion in Ontario consumes natural resource lands and prime farmland (Land Directorate, 1985) but the remaining farm land is being used more intensively. As land and buildings have risen in value, the number of farms continues to diminish. Between 1986 and 2006 the number of Ontario farms declined from 72,713 to 57,211 (Statistics Canada, 2008-10-31) as the average farm size has increased (Putnam, 1959; McCuaig and Manning, 1982; Dey, 2008).

SUSTAINABLE DEVELOPMENT AND ONTARIO AGRICULTURE

One of the criticisms that the Brundtland Commission's path-breaking report, *Our Common Future* (1987), made of the North was that its agriculture was subsidized and this damaged Southern agriculture. This is a theme Mazoyer and Roudart (2006) returned to when they argued that our relatively low prices of food prices are partly due to this subsidization. Unable to produce everyday subsistence food like maize, rice and wheat as cheaply as the North, many peasant farmers have been forced to leave their rural areas for occasional employment in the cities.

The Brundtland Commission's alternative development paradigm has a great deal to say about the relation of population to natural resources and the limits to economic growth but for our purposes, the key areas that the Commission addressed were food security, the loss of genetic resources, biodiversity and energy. The Commission argued that there must be ways of dealing with the environmental problems connected with intensive agriculture, so that subsistence and small operation farmers can be protected as the environment is preserved. The Commission also advocated the search for and development of alternative forms of energy so that production could be less energy intensive.

Still, *sustainable development* remains a contested concept with respect to how people would like to manage the direction of social change. The form that this debate takes depends upon whether people support an ideal state of sustainable development, a strong yet achievable form, weak sustainability partly at the cost of economic growth, mere pollution control with virtually unbridled growth or the rejection of the concept sustainable development concept entirely, as the deep, ecocentric ecologists have. The latter see it as an anthropocentric attempt to manage nature itself instead of seeing nature as intrinsically valuable so that a reciprocal partnership between nature and humans can be developed. Whereas those favoring strong sustainable development consider environmental protection as a precondition for economic development, advocates for a weaker form of sustainability claim instead that economic development is a precondition for environmental protection.

At the time of Deloitte's Dec. 2007 glowing review of the effects of the work on the Ontario Ministry of Agriculture, Food and Rural Affairs (OMAFRA) contract with the University of Guelph on the economy and the provincial environment, a leader in the Ontario Farm Environmental Coalition (OFEC) remarked that, thanks to the widespread use of no-till or low-till cultivation, the reduced use of pesticides and herbicides and the growing use of riparian buffer strips between crops and streams, the health of our agro-ecosystem has never been better (Dec. 18, 2007).

In sharp contrast, Monpetit's book, *Misplaced Distrust* argues instead that the creation of OFEC in the aftermath of the 1990 election of the New Democratic Party (NDP) was part of a farming coalition's attempt to escape from the effects of the Environmental Bill of Rights promulgated by the NDP.

Monpetit believes that the ensuing Farm Registration and Farm Organizations' Funding Act gave the Ontario Federation of Agriculture and the smaller Christian Farmers' Federation of Ontario enough clout and resources "to subordinate OMAFRA in a clientelist network within the agro-environmental actor constellation. The ministry is reduced to providing services of minor strategic importance within the framework set by the Environmental Farm Plan" (2003: 102). This book's case studies will help to show that while agriculture's negative environmental 'externalities' continue to happen, it is no longer fair to describe OMAFRA's relationship with farm organizations as merely that of patron and clients.

AGRARIAN POLITICAL ECONOMY AND DIALECTICAL SYSTEMS THINKING

In order to understand the contradictions and complimentarieties between the big systems like the international food regimes and local governance and farming systems, this book utilizes a combination of agrarian political economic and dialectical systems thinking. Though the two are closely linked, I will begin by outlining the perspective provided by the former.

Saskatchewan political economist/ecologist John Warnock (1987) has been a strong opponent of the increasing displacement of family farm agriculture with capitalist industrial agriculture and corporate control of food chains internationally. He complains that industrialized capitalist agriculture produces food at a huge fossil fuel subsidy taking about three calories on average of energy for every calorie of food it produces. This form of agriculture does not produce food for the local market but instead allows corporations to control the manufacture and distribution of food for the international market.[2] Farmers are caught up in an increasingly contradictory position between their economic viability and environmental protection as well as between large multinational agri-business and pharmaceutical companies and ecosystem health.

After losing an expensive legal case with Monsanto, one angry American farmer complained that

---

[2] Poultry and dairy farmers are, of course, protected by supply management in Canada which at times leads to higher prices for consumers. They are not necessarily as competitive internationally as they might otherwise be (see Pfeiffer and Filson, 2004) but many countries enjoy one form or other of supply managed dairy and American dairy farmers have benefited from large subsidies as well.

"In the case of Monsanto, their control is so dominant. If you want to be in production agriculture, you're going to be in bed with Monsanto. They own the soya bean. They're going to control that product from the seed to the supermarket. They are in fact gaining control of food" (Kenner, 2009).

The same kinds of pressures affect farmers in Ontario and help to explain the dominance of soybeans, corn and livestock and the relative scarcity of nutritious vegetable production, except in Ontario's Holland Marsh which exports more than half of its product.

In his review of agrarian political economy, Buttel (2001: 175) also highlights the fact that the international agri-food system has become increasingly standardized and homogenized but he also identifies such emerging "countertrends—toward organic food, local food systems, local food labeling, an emphasis on 'quality,' and so on—that are leading to widespread restructuring of regulatory practices along the entire span of food chains". While there is growing interest in local food as well as the fact that organic production cannot grow fast enough to meet the demand for organic food products, the latter continue to represent only a few percent of the total number of producers and local food is no panacea for the continuing farm financial crisis (Filson, 2004b).

Unlike Cornell's Pimentel (2008; Pimentel et al., 2000) who has also critiqued capitalist style conventional agriculture but specifically equates sustainable agriculture with organic agriculture, the focus of this book is on the majority of Ontario farmers, only a small minority of whom produce food as certified organic farmers. On the other hand, Warnock (2002) believes that Pimentel's criticism of monocultural corn production is well taken due to its destructive soil influences and excessive fertilizer, pesticide and fossil fuel dependence. But while this critique of monocultural commercial farming arises from a political economic/ecological perspective, our political economic focus is instead on farmers' environmental management systems and the adequacy of those systems.

Within capitalist countries, governments tend to react to the changes that take place within the productive and distributive realms seeking to protect the conditions for accumulation while legitimating the prevailing relations (O'Connor, 1994). They can do this within the agri-environmental realm by promoting a stewardship ethic and by *regulating* the most egregious behaviour in order to better protect the environment. The 2007-2009 industrialized capitalist recession[3], on the other hand, was caused by a *lack of financial regulation* and not just the glut of housing production occasioned by the availability of sub-prime American mortgages but also by a glut of automobiles and other forms of production relative to effective demand. As the prosperous period moved toward crisis and a credit crunch there was a continuing period of bogus prosperity when prices, including for agricultural commodities and food, rose independently of their value. Of course our relatively cheap food, including so-called junk food, is actually very expensive because of the health and environmental costs of the agr-food system.

Though some food prices have declined during the recession, the international food crisis is expected to continue as the industrialized economies slowly rebound both because of rising demand for food from emerging economies like China and India as well as the growing use of corn, sugar cane and related foodstuffs to produce ethanol based fuels.

Leaving these contextual issues aside for now and returning to the book's agriculture/environmental security focus, it should be noted that most of the research upon which the book is based utilized data collected from interviews, focus groups and mailed questionnaires with farmers, non-farm rural people, conservation authority personnel and others within related agricultural institutions in Ontario. Its theoretical perspective draws upon dialectical systems thinking of the complex relationships among farming and ecosystems, food production and consumption, commodity and non-commodity agricultural outputs.

If we use contradiction to mean that there are two or more processes related to each other which both support and undermine each other (Ollman, 1993) we can consider farming systems as well as agri-food systems in the more general sense to contain both internal and external contradictions with other systems. These systems are greater than the sum of their parts and they exist in a continual state of flux as a result of their internal and external contradictions. Harvey (1996: 8) elaborates:

---

[3] As capitalism becomes more concentrated and centralized, which has happened not only with agricultural production in Ontario but with many other industries as well, the potential developed for rapid increases in accumulation and, consequently, relative over-production of so-called 'fixed capital' in plant and equipment, on the one hand, and relative under-production of 'circulating capital' (the demand for organic raw materials like fuel growing more quickly than their supply) on the other. Since raw materials are one of the main elements of constant (which includes fixed and circulating) capital, low prices for them are crucial for the industrialized capitalist countries. This is because the aliquot part passing into the product due to depreciation of machinery steadily decreases while the value of circulating capital passes entirely into the product (Marx, 1959).

"To say that parts and wholes are mutually constitutive of each other is to say much more than that there is a feedback loop between them. In the process of capturing the powers that reside in those ecological and economic systems which are relevant to me, I actively reconstitute or transform them within myself even before I project them back to reconstitute or transform the system from which those powers were initially derived."

Harvey argues that the constantly changing nature of systems is probably the most important dialectical principle to grasp.

Dialectics is a way of understanding the flows of capital (productive labour, fixed and circulating capital, goods and money). Capital is simultaneously a process involving the circulation of capital as well as a stock of assets. From a dialectical/systems perspective it is important to study the transformative moment within the realms of production, for instance, to determine how and why peasant and simple commodity producing farming systems become transformed into capitalist systems with greater investments in fixed and circulating forms of capital than labour. While it is clearly too simplistic to say that the dialectical analysis of contradictions—between farmers and nature, between farmers and farm workers, between rural and urban, between cheap food policy and expensive health care, etc.—can simply be understood as a movement from thesis, to antithesis and eventually to a new synthesis, this is a useful heuristic for understanding the dialectics of how these agrarian and other productive systems change. Stable systems are therefore not the norm so it is essential to grasp the ways in which the systems we are studying are changing as the result of their internal and external contradictions and complementarities (Harvey, 1996).

Ecosystem health practitioners use similar language to describe systems thinking and complexity (see chapter two) These complex systems are comprised of feedback loops connecting the parts of the systems which are externally linked to other systems. These parts are organized hierarchically and contain nested hierarchies within their multiple levels so it is essential to recognize the connections existing between food systems, farming systems, health systems, ecosystems because they each affect one another, transcend borders and are tied in with the dominant and residual systems of production within the international economy (Nuedoerffer *et al.* 2005).

FARM CLASS STRUCTURES

Rather than viewing social class in the gradational terms of upper, middle and lower, class is herein described in the discrete sense of one's relationship to the means of production. Those owning and controlling the means of production are *capitalists* while those who sell their labour for a wage are *working class*. A small farm operator without employees usually works off the farm and is therefore a *worker farmer*. Farmers who employ a few farm hands from time to time are *petty bourgeois*. Other classes exist and are described in terms of whether or not they control the means of production as do managers and supervisors (Wright, 1978). These are not the class terms in which farmers view themselves but from a political economic and dialectical systems perspective it, it makes good analytical sense to construct their classes in these discrete terms.

Thus these farming classes include the capitalists, simple commodity producing small family farm operators (petit bourgeois), farm workers including the migrant working class and, of course, the worker/farmers. The separate and distinct class locations of different types of farmers and farmers producing different types of commodities affects their class consciousness, their mindscapes[4], their attitudes toward farm organizations[5] and of course their attitudes to the legitimate prospect of unionization for Ontario's 32,000+ farm workers. Some class fraction of farmers practice environmentally friendly forms of farm management and others do not, with many in-between, in large part because of the quality of their land, the size of their inheritance, the way farmers are socialized including ethnically and religiously as well as how open they are to learning how to adapt to the use of new technologies including environmental best management practices.

Recent work by Sparling and Laughland (2006) has pointed to the "two faces" of Ontario farming: a small group of large producers and a large group of small producers. Sparling (2007) observes that the so-called 'decline of

---

[4] The ways in which farmers, professionals and others perceive the agricultural and non-agricultural products, assets and flows within landscapes constitute their evolving mindscapes. See chapter 2.

[5] These organizations include the National Farmers' Union (small to medium sized family farmers), the Dairy Farmers of Ontario (DFO) (close adherence to their supply managed organization regardless of class location) or the more conservation commodity based agricultural organizations (e.g. Corn Producers, Cattlemen's Association, Soya Producers, Ontario Pork, etc. and their meta organizations Ontario Federation of Agriculture (OFA), the Christian Farmers Federation (CFFO) and collections of organizations like the Ontario Farm Environmental Coalition (OFEC) and the Ontario Farm Animal Council (OFAC).

agriculture' is actually only the decline of small farming operations. Off-farm employment now exceeds net farm income, with the exception of dairy, for the other main farm sectors of grains and oilseeds, hogs, beef and greenhouses. This is particularly apparent in Ontario's grains and oilseeds and beef sectors where small farms predominate. Incomes within the dairy sector are stable (as they are with supply managed poultry) with the average dairy farm's equity being $1.9 million and incomes are reasonably good thanks mainly to supply management. Hog producers, while less stable, have had medium sized revenues as well though the perfect storm of excess supply and H1N1 (swine) flu have had the same impact on hog production as BSE[6] (mad cow disease) had on beef earlier. Growth within Ontario's greenhouse sector is very strong above the $250,000 revenue level with the average having revenues per year of just less than $1 million whereas smaller greenhouse operations are declining. There are fewer small dairy and hog operations and an increase in larger farms even though the total numbers have declined. On average, the necessary scale required for farm survival is therefore increasing (Sparling, 2007).

Though most agricultural produce in Ontario is produced by large, commercial farms (Sparling and Laughland, 2006), most Ontario farms remain small, family operated farms. While the large farms are often still family owned farms, they are also often capitalist enterprises employing farm labour and producing products for export with the help of some hired labour and/or the occasional use of custom labour for everything from haying to harvesting.

Cooperatives also offer an alternative whereby a group of relatively small farm operators can combine their efforts to obtain many of the technical advantages of large scale capitalist farming with the social benefits of family farming. Ontario agriculture therefore includes small producers with off farm employment, cooperatives and larger scale, capitalist farms where workers and managers are employed by farm owner operators. In November, 2008 Ontario's Court of Appeal has allowed Ontario's farm workers to certify as unions leaving Alberta as the only province which does not allow farm workers to unionize (Preibisch, Nov. 25, 2008). Because farmers exist in several different class categories, different agro-environmental policies are needed for different types of farms of varying farm sizes.

---

[6] Cases of bovine spongiform encephalopathy (BSE) began to be discovered in May, 2003 and until recently this has cast a pall over the Canadian beef industry (Goddard, E. and Unterschultz, J. 2004).

Assessing Agro-Environmental Behaviors and Policies by Studying Watersheds

Even though fixed boundaries within the agro-ecological system are somewhat arbitrary, for the purpose of our research we focused on watershed ecosystems. The farming systems, such as the various livestock and plant systems exist within agro-ecosystems so choosing to conduct research with farmers and other actors within specific southern Ontario watersheds containing most Ontario farms provided a convenient staging area for this research. Figure 1.1 provides a map of the study areas which are included in this book with the black dots showing the five watersheds of primary interest in the chapters below from southwestern to eastern Ontario: Ausable Bayfield, Maitland Valley, Grand River, Lake Simcoe and Raison Region.

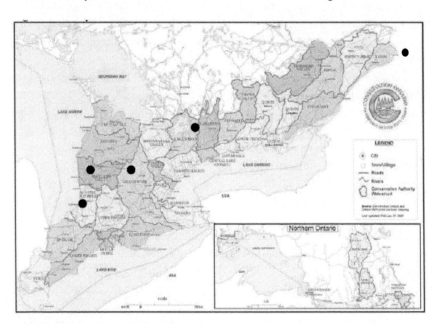

Figure 1.1: Study Areas: Rural Ontario Conservation Ontario, 2005
(www.conservation-ontario.on.ca/find/index.html)

These sample watersheds were selected based on their agricultural importance and their representativeness of southern Ontario. The environmental ethics that underscores our research is based on the notion that environmental security requires an understanding of the centrality of farmers' attitudes and motivations which affect their environmental interactions including ecosystem functions, structure and overall sustainability.

Another reason for selecting watersheds is because each of Ontario's 36 watersheds are self-contained, interconnected ecosystems. In 1946 the Conservation Authorities Act was created to address the need for integrated management of Ontario's watersheds by what is known as Conservation Authorities (CA's) (Saugeen Conservation, 2008). These CA's are ultimately responsible for the stewardship of environmental resources within the specific watershed for which the particular CA is responsible. The CA's work with municipalities in an attempt to balance the various needs for urban and agricultural development with recreation and wildlife habitat needs. In part they do this by working with landowners to encourage the adoption of environmental best management practices. In agriculture, the CA's work with the Ontario Soil and Crop Improvement Association (OSCIA) to promote the implementation of EFP and cost share programs that the CA's have to encourage soil, water, air and biodiversity conservation such as the Clean Water Programs provided by the following CAs: Ausable Bayfield, Catfish Creek, Grand River, Kettle Creek, Long Point Region, Lower Thames Valley, Maitland Valley, St. Clair Region and the Upper Thames River. As OSCIA program manager Andrew Graham argues "Conservation Authorities have done a great job of helping people think beyond farm boundaries and look at the entire watershed of which they are a part" (Protecting Ontario's Water, June 27, 2007).

BOOK OVERVIEW

This book opens with a discussion of the global food and agrarian crises which most recently have coincided with a generalized international capitalist crisis of production and exchange. Before focusing on the ways in which intensifying Ontario agriculture impacts the environment, this chapter discussed the wider international context of the relationship between food and agrarian crises, globalized food regimes and how the relentless transformation of farming impacts food prices, generates intensified commercial farming and to some extent threatens environmental security. The most important lens which we are using to understand prominent changes in agri-food systems is agrarian political economy and a form of dialectical systems analysis of environmental management within agro-ecosystem watersheds. This requires an appreciation of the class character of farming including the strengthening economic position of capitalist farmers who own their own means of production, control investment and employ the most vulnerable, often migrant, farm workers. The contradiction between these latter two classes became evident once again when Rol-Land Farms summarily dismissed 70 farm workers in the aftermath of the Appeal Court ruling giving farm workers the right to unionize. Another weaker group of agriculturalists includes the small farm operations run by worker farmers with off farm income who still constitute the majority of Ontario farmers despite their marginalization.

Chapter two discusses environmental security and agro-ecosystem complexity. Landscapes and mindscapes are introduced prior to a discussion of multifunctional agriculture and our poly-ocular advocacy for strong sustainability. By contrast, weak sustainability advocates want to allow greater market freedom and place a higher degree of confidence in future technological changes which they feel will cope effectively with environmental problems like excess greenhouse gas production (Baker, 2006). An underlying theme of this book, developed through the use of case studies of farming in southern Ontario watersheds in later chapters, is a critique of weak sustainability policies and behavior in the agriculture sector and the provision of a rationale for why stronger forms of sustainability are needed. Chapter two concludes by introducing some agro-environmental policies, discussed at greater length in chapter three.

Improved environmental security requires policies which take account of how farm size and structure as well as farmers' mindscapes affect their environmental interactions with agro-ecosystems and overall sustainability. In the absence of this knowledge and sufficient funding, plans to protect ecosystem sustainability and resilience have often been inadequate. Because agro-ecosystems are complex and contain implicit uncertainties, protecting our environmental security from the excesses of intensive agriculture has been a major challenge. Our research therefore utilized a statistical and qualitative data analysis of extensive interviews and surveys of farmers, conservation authorities, and non-farm rural people's views within southern Ontario's major watersheds to better understand the situation. Provincial and federal legislation in addition to comparable policies abroad, both of a regulatory and voluntary nature, are reviewed and then compared with what exists in the European Union (EU) and the United States in chapter three. We acknowledge the complexity of the issues and the fact that in some respects we have much to learn from other jurisdictions. The importance of Ontario's Environmental Farm Plan (EFP) is explained. The barriers to farmers' participation in agro-environmental programs, the importance of technical and financial incentives and the need to understand the differences as much as the similarities among farmers in the design of appropriate environmental programs are reviewed. Based on the study of these jurisdictions we conclude that environmental management systems (EMS) and the adoption of 'best management practices' (BMPs) have the potential to strengthen environmental resilience.

Case studies of farmers' environmental management within the Ausable Bayfield, Maitland Valley, Grand River, Lake Simcoe and Raison Region watersheds of southern Ontario are then presented in chapters four to six in order to reveal the major factors underlying this adoption behavior so that more suitable environmental and social policies can be proposed in the final chapter. Chapter four looks at two sub-watersheds within the Grand River, Ontario's largest watershed, which drains into Lake Erie. The chapter presents a study of the farmers' environmental attitudes, behavior and perceived quality of life based on an analysis of focus group meetings and completed questionnaires by farmers and non-farmers in the most degraded sub-watershed and the least degraded sub-watershed. The environmental attitudes and practices of farmers within an environmentally problematic sub-watershed of the Grand River is also analyzed. The widely held view of the inherent sustainability of these mostly small, mixed farming operations is partially upheld by the fact that many of them are genuinely organic farmers. On the other hand, the environmental problems in this sub-watershed and the farmers' relatively low level of awareness of and participation within Government and Conservation Authority environmental programs is probably connected with their fairly low average level of formal education and the intensity of the number of livestock operations in the area.

The Ausable Bayfield watershed, which drains into Lake Huron, is the focus of chapter five. That chapter contrasts the factors that motivated farmers to adopt BMPs in a relatively healthy sub-watershed with the behavior of farmers in a problematic sub-watershed. Furthermore, it provides insight into landowners' perceptions regarding voluntary, collaborative and regulatory programs recently implemented. It also considers many of the issues that are of interest to environmental officials planning to improve environmental management programs and policies.

The sixth chapter analyzes hypotheses developed from the earlier case studies to evaluate the latest perceptions of farmers within the Ausable Bayfield, Grand and Lake Simcoe watersheds. It explains the importance of farm size, gross farm sales, gender, implementation of an EFP and a nutrient management plan (NMP), full time farming and relatively large livestock operations on the ability to predict higher BMP adoption rates. It also points to the importance of developing environmental programs to find ways to target and influence small farmers to adopt environmental BMPs as current programs including the farm safety net have generally been more effective on larger farms.

Thus, the empirical findings from these studies support the advocacy of a stronger form of sustainability than is presently practiced to protect both the environment and the farming community. The lessons learned from these studies range from the importance of taking a watershed management approach to the development of clean surface and ground water and the education of people within each watershed. There must be a continuous review of programming to develop collaborative plans at the individual and watershed levels before these plans are implemented. Different responsibilities are identified for the federal, provincial level, the conservation authority/ municipality level and the farm level. The cost of programming must be shared and include a variety of conservation tools, risk management and accommodation of urban, industrial and agricultural growth. Different strategies are also required for at least the variety of different farming sectors and especially large, commercial farmers versus small producers who depend heavily on off-farm employment. This book concludes with the recognition of the importance of moving toward the recognition and development of multifunctional agriculture where farmers produce both food and fiber and also environmental goods and services.

In: Agricultural and Environmental Security...   ISBN: 978-1-61668-156-2
Editor: Glen Filson   ©2011 Nova Science Publishers, Inc.

Chapter 2

MULTIFUNCTIONALITY, ENVIRONMENTAL SECURITY AND STRONG SUSTAINABILITY

Katia Marzall, Glen Filson and Bamidele Adekunle

INTRODUCTION

This chapter discusses the elements of a political economic analysis of the environmental consequences of how differing farming classes with varying sizes and commodity types manage their impacts on some major watersheds in southern Ontario. Before looking more closely at how these farmers manage environmentally, within this chapter we establish some of common requirements for environmental security within an increasingly multifunctional set of landscapes and farming mindscapes. Farmers' ethno-cultural characteristics combine with farm structure, the leadership role of various farm organizations as well as farmers' social class locations in determining the extent of farmers' other environmental behavior. All of these factors affect farmers' motivation and the intentions to act in environmentally responsible ways (Garling, 2003). Farmers will adopt best management practices and act to reduce the environmental risks of their operations if their awareness, organizational, financial and technical supports enable them to act in environmentally responsible ways. Unfortunately for many, these conditions are relatively unfulfilled.

Some critics (e.g. Warnock, 2003; Imentel, 2008) argue that as intensive capitalist agriculture is increasingly concentrated on larger and larger farms which are highly specialized and may house as many as 10,000 pigs, a few thousand cattle and several thousand hectares of farmland. This has a big impact on the environment through various pathogens, phosphates and nitrates which have had an increasingly adverse affect on our watersheds, especially where there are concentrations of intensive livestock which exceed the amount of land farmers have on which to spread the resulting nutrients.

While Ontario imports about 80% of the fruit and vegetables it consumes, almost half of its cropland is devoted to soybeans and corn (2.4 million and 2 million acres in 2008) (Reinhart, 2009 and Webb, 2009). These two crops go into the production of many things besides food, including industrial products, pharmaceuticals, ethanol and animal feed. "A bushel of corn produces some 440 two-once bags of 99-cent [Dorito] chips. [The] farmer grosses $3.70 for the bushel of corn, Doritos more than $440" (Webb, 2009) Increasingly Ontario possesses farms with as many as 2,000 to 3,000 hectares of land (Webb, 2009), however, it's also true that a myriad of small, simple commodity producing farms concentrated in significant numbers within a watershed can have chronically negative environmental consequences if the farm operators lack the time and means to be more environmentally responsible (see chapter four).

ENVIRONMENTAL SECURITY, COMPLEXITY AND TRANSFORMING AGRO-ECOSYSTEMS

The issue of environmental security is a current concern for many because of the growing awareness of the influence humans have on the sustainability of their ecosystems. Environmental security is related to the degree of environmental threats to human life (Conca and Dabelko, 2004; Nef, 1999; Falkenmark, 2001; Klubnikin and Causey, 2003) and is critical to agricultural production and to the sustainability of rural communities. Dalby (2000), Homer-Dixon (1999) and Gleick (1991) have written about how an unbalanced state of the environment can lead to social unrest, various types of conflict and even war. According to Lonergan (2000: 67-68), there are four categories of interpretation for the concept of environmental security:

1. Security of the environment (or security of services provided by the natural environment: this has also been interpreted as non-diminishing capital);

2. Environmental degradation and resource depletion as potential causes of violent conflict;
3. Environmental degradation and resource depletion as threats to national welfare (and, therefore, national security); and
4. Environmental degradation and resource depletion as two of many integrated factors which affect 'human security'.

The Foundation for Environmental Security and Sustainability argues that political and economic stability as well as people's well-being are crucially dependent on environmental security. They observe that the link between the environment and security is very visible in the developing world where people are food insecure in large measure because of environmental devastation caused by extractive industries, water and forest conflicts as well as energy shortages (ESSS, 2008).

In the agricultural context, there is concern regarding the high dependence of and the strong influence on environmental conditions. The environmental security of agro-ecosystems must be considered in order to promote the sustainability of Ontario rural communities wherever agriculture is practiced. Environmental security is also obtained when people feel responsible for their environment. Responsibility can arise because of farmers' morality and social expectations (Kaiser and Shimoda, 1999).

The complexity of ecosystems' structure and processes, as well as their intrinsic uncertainty, only increases the challenge to find effective approaches to enhance environmental security. In order to understand which interactions promote environmental security and which are detrimental, requiring redefinition, we face the challenge of understanding the drivers of environmental security. To analyze environmental security we need to consider the complexity of the interaction processes between farmers, their non-farm rural neighbors and their environments. The ways in which farmers, professionals and others perceive the agricultural and non-agricultural products, assets and flows within landscapes constitute their evolving mindscapes. This analysis can only be successful if there is a proper understanding of the logic that drives landowners' interactions in and with the environment and what shapes the agricultural landscape (Marzall, 2006).

This challenge requires the combination of ecological, sociological, anthropological and psychological knowledge, among others, which are usually not considered in a holistic manner (Berkes, Colding and Folke, 2003). It requires a framework that can be easily applied to real situations. This book undertakes this challenge, exploring the most important drivers in the opinions of government officials, academics and especially farmers' environmental perspectives, observing their interactions, proposing an analytical framework and considering its feasibility in southern Ontario watersheds. In part, the authors used an ethno-environmental approach, where insiders are involved in enunciating the best environmental management processes and policies that can create incentives for adoption. The epistemological premise of this framework is that the inherent complexity, which characterizes environmental security, can only be effectively understood from an equally complex yet intelligible and comprehensive conceptualization (Nef, 1999). Some degree of complex, interdisciplinary and dialectical systems thinking is required which allows for the integration of biophysical and social-cultural knowledge realms.

The acknowledgment of complexity and our limited view-points provokes caution about what we say regarding our observations of the environment. Complexity and non-linearity go hand-in-hand with uncertainty (Berkes, Colding and Folke, 2003). After the earth was formed, it did not remain a mass of molten lava, but conditions evolved which allowed for the development of plants and animals. Change often implies an unknown future and an understanding of it acknowledges surprises (Gallopin et al, 2001). Therefore, our major challenge is not to avoid change and uncertainty but to learn how to live with change and adapt (Funtowicz and Ravetz, 1994).

The challenge of embracing uncertainty is also found in the daily life of farmers. This challenge is greater in the processes followed by agricultural scientists who develop and (try to) establish the rules that farmers should follow, as with the cross-compliant rules presented below. These rules of apparent stability only emphasize our distance from perceiving the environment's dynamics, disabling accurate recognition of imminent changes, requiring our ingenuity to acknowledge contradictions and adapt. Ravetz (1993) stresses the importance of the awareness of knowledge's uncertainty. The ignorance of ignorance, he notes, can lead to errors, and often disasters. The belief in the ultimate truth of science, for instance, might dangerously lead to paths believed to be the solution for problems, but instead create new ones. With the purpose of increasing food production, scientists followed some paths of certainty that led to the consequences stressed by Rachel Carson in her 1962 path breaking book *Silent Spring*, and for many years following (for example, soil erosion, water contamination, loss of biodiversity and food contamination).

In order to take the issue of environmental security seriously we have to reflect on our motivations and on the outcomes that our actions generate. We need to ask what we actually want from the environment and what we are accomplishing through this quest. The promotion of environmental security is the objective of environmental extension (Marzall, 2003). Its aim is to promote reflection on the impact that our actions impose on the environment, and to consider alternative interaction processes. While environmental extension processes might guide policy development and help re-structure institutions involved with environmental management and research, at this point we are concerned with individual daily decisions. More specifically, we are concerned with farmers' responsibility for their environmental behavior and the changes they effect on the environment in the process of food production as well as the production of non-agricultural goods and services. Farmers' environmental responsibility is also conditioned by the types of commodities they produce, their farm structures, types and farm sizes (Filson, 1993, 1996).

Recent works, such as Berkes and Folke's (1998), Colding and Folke's (2003) and Filson (2004) use systems perspectives to observe the linkages between humans and the environment and to combine a variety of components that contribute to these interactions. A dialectical systems approach allows for a view of the different components of a system, as well as of the interactions among them including their contradictory and complementary elements. We cannot think reasonably about the environmental impacts of agriculture without grasping the extent to which the global food regime within which farmers function, has conditioned them to produce a few commodities—corn, soybeans, wheat, beef, dairy, poultry and pork—at the behest of a few large private "food processors, retailers and agricultural businesses [that] are generating massive profits delivering cheap food by squeezing farmers' incomes, forcing farmers into environmentaly unsustainable practices—or out of business" (Webb, 2009, A11). The extent to which the farmers have become absorbed into production for the global, capitalist food regime explains why so few farmers are able to produce for their local market.

Dave Ferguson, a farmer who produces corn and soybeans near the relatively polluted Sydenham River which drains into Lake Huron explains that the demand for these crops "combined with competition from ever cheaper global imports, has placed relentless pressure on farmers not only to grow these crops, but to expand" (Webb, 2009). This cheap food policy also helps to explain why Ontario farmers do not produce the vegetables and fruit demanded by our increasingly multicultural province and why so much of our food has become junk food which has serious negative consequences for our health and environment. Thus contradictions between our global food system, people's health and the environment are creating environmental and health costs including pollution, obesity and diabetes as the result of the squandering of fossil fuels in agricultural production and the spewing of greenhouse gases, excess nitrates, phosphates and glyphosate into our watersheds.

Too much of our food system is based on monocultural food production which caters to food processors who produce high calorie foods with lower nutritional and that there is not enough support for a diverse agriculture which produces many of the fruits and vegetables as well as existing mainstream foodstuffs that are particularly desired by our increasingly multicultural population. The extent to which we are integrated in the latest food regime and are controlled by an industrialized capitalist system of production and processing often jeopardizes local control by farmers and rural communities and is too fossil fuel dependent to benefit the environment. This leads to other contradictions between our cheap food system and the expensive health care consequences resulting from the obesity, creation of type II diabetes, hypertension and even cancerous consequences of many of our foods. Usually the relatively wealthy can afford nutritious foods while the working class and surplus population is virtually forced to eat the junk food that so damages their health (Wallinga, 2009; Muller et al.2009).

Environmental security is related to the idea of resilience (Berkes and Folke 1998; Adger, 2000). Resilience is the capacity of a system to continue to evolve throughout its dynamic life cycle, despite internal and external stresses. It expresses a system's capacity to cope with uncertainty. The major feature of a resilient agro-ecosystem system is the capacity farmers have to perceive options, and to be able to choose those options that will promote environmental security rather than undermine it (Marzall, 2006). The changes the farmer imposes on his/her environment adds to, changes or destroys elements of the ecosystem. If the elements of that structure of the resilience are lost in this process, resilience is also lost, or undermined. In this way, the environment becomes more vulnerable.

There are two major components that constitute the resilient capacity of a system, therefore of environmental security: actual and potential security (Marzall, 2006). Actual security is defined by the environmental conditions that allow for security and it encompasses the environmental elements that express security, including those that allow for life quality, supply ecological services and health. These structures can be observed at the landscape level. If there is sufficient biodiversity, air quality, soil and water, and synergistic connectivity among these different elements so that agriculture can produce safe, high quality food and fiber and perform rejuvenating ecological environmental services within that agro-ecosystem, it is environmentally secure as well as resilient. Human security depends upon environmental security but it also requires economic, social, political and cultural security (Nef, 1999).

An example of environmental insecurity occurred during Ontario's Walkerton tragedy as the result of a pathogen originating from a nearby beef herd in the year 2000. People's actual environmental security was compromised when, as the result of taking for granted the safety of their drinking water despite the unknown presence of *Escherichia coli* 0157: H7 bacteria, seven people died and more than 2,100 people's health was permanently damaged. Poorly trained, incompetent local officials jeopardized local people's environmental security by failing to treat the water with sufficient chlorine to kill the pathogen. This case of environmental insecurity has been blamed on the Ontario Conservative Government under Premier Mike Harris which made significant cuts to regulatory environmental spending (Whittington, 2008) by as much as one-third (Montpetit, 2003).

Potential security involves human perception of security and capacity to deal with uncertainty on the one hand, and, on the other hand, it includes the influence humans have in shaping their environment, influencing its resilience. Dealing with uncertainty involves being able to find options, when unexpected events arise, especially if the event has negative impacts. On the other hand, behavior that will not compromise actual environmental security is essential. Humans are part of the ecosystem. If they change the environment to an extent that this environment becomes more vulnerable to external stresses, they are compromising the environment's resilience and threatening their own security. A depleted environment will have fewer options for society to choose from, when faced with further impacts or stresses (Marzall, 2006).

Yet as will be seen in chapters 4-6 below, many small, precarious farm operators see this from the other end. They place their financial security ahead of the environment feeling that until they can gain a financial foothold to reinvest, they cannot afford to implement expensive environmental best management practices.

Lonergan's (2000) view is that we need to consider the environment and security from an integrative, interdisciplinary perspective. Only then will we be able to understand environmental security. The issues at stake in Nef's (1999) view relate much more to the overarching issues of sustainability, survival, well-being, quality of life and human development, transcending the political realm (Marzall, 2006). To do this requires a dialectical perspective which understands how the constantly changing, internally and externally contradictory social and ecosystems evolve.

Together, these two concepts define environmental security. As scholars, policy makers, extension agents or researchers, our interest is in potential security. We want to learn about how human action impacts the environment, what potential alternatives exist to these actions, and how these alternative actions can be promoted.

Environmental concerns encompass food production, natural disasters, water and energy supply and are among other issues that deal with our biological life. Cultural and economic issues are also affected by these environmental concerns yet often become of secondary importance due to the urgency to guarantee that basic human needs are met. Based on these environmental concerns, it becomes almost natural to talk about 'environmental security' as the necessary structure that will provide human security (Lonergan, 2000). Environmental concern plays an important role in behavior changes and it leads to environmentally responsible behavior (Fransson and Garling, 1999). Knowledge and norms also affect the way environmental concern leads to environmentally responsible behavior.

We are definitely more aware of environmental problems than ever before. Many people have a greater sense of the degree to which environmental problems and concerns exist in many parts of the world but it is also the case that the magnitude of environmental impacts of human behavior is more substantial now than earlier (Redman, 1999; Berkes, Colding and Folke, 2003). Moreover, we are more aware that many of the unwanted changes that have occurred including such environmental consequences of human behavior as pollution, water depletion and climate change. We are more aware that "...facts are uncertain, values are in dispute, stakes are high, and decisions are urgent" (Funtowicz and Ravetz, 1994: 1882).

Farmers are aware that their cultivation methods affect soil structure and fertility as well as the quality and quantity of water in the rivers within their region, and even the availability of fish in these waters. The choice of farming methods has great implications on the shape of the landscape and the quality of the environment. Dense forests have sometimes been converted into fields, deserts have become orchards, and rivers have become deserts through the activities and technologies adopted by farmers since the beginning of agricultural times at least 8,000 years ago (Mazoyer and Roudart, 2002).

Yet, as farmer Dave Ferguson reminds us, farmers cannot be expected to care for the land they farm without the help of society. "Until society gets in their mind that they have to pay to get these farms sustainable...If you want cheap food, that's what you're going to have" (quoted in Webb, 2009, A11).

In agro-ecosystems, rivers that are not filled with soil sediment carried by erosive processes, or with organic and inorganic pollutants, are much more likely to maintain their water capacity and life structure in occasions of drought, for instance, than contaminated rivers. In the case of excessive rains, the structure of the riparian ecosystem and the extent of human presence in proximity to rivers define the degree of flood damage, rather than the amount of rainfall exclusively (Frank and Vibrans, 2003). Other examples are riparian areas, or native vegetation patches maintained to host local birds, mammals, insects and reptiles, which would otherwise transfer their homes to the agricultural fields, becoming pests instead of local fauna (Nilsson and Svedmar, 2002).

The loss of habitat is one of the most severe problems faced by southern Ontario (Neave *et al.*, 2000; Filson, 2004a). Urban sprawl and agricultural practices including the draining of wetlands, the use of inorganic fertilizers and pesticides all threaten wildlife but Neave *et al.* (2000) argue that converting wildlife habitat into farmland is the biggest problem of all for the Mixed-wood Plains region of southwestern Ontario. Agricultural practices affect ground water through increasing the risks of bacterial contamination arising from livestock production as well as excess nitrogen and phosphorus from organic and inorganic fertilizer. Similarly agricultural pollution affects surface water from fertilizers and pesticides. With the growth of intensive farming systems, significant water and other pollution problems have arisen (Samson *et al.*, 1992).

Canadian agriculture has contributed to climate change through the production of greenhouse gases such as methane, nitrous oxide and carbon dioxide but while this effect has been increasing in the prairies it has been relatively stable in Ontario. Soil management has steadily improved with less conventional tillage and less bare soil allowed though soil compaction has increased in some parts of southern Ontario. Energy input has increased somewhat while energy output has fallen, mainly due to the growth of intensive farming in the region (Filson, 2004a). In order to continuously improve agro-ecosystem management and environmental security, it is important to understand how landscapes and mindscapes interact.

LANDSCAPES AND MINDSCAPES

Considering landscapes in relation to the mindscapes of the humans inhabiting them is one lens which may be used in order to observe the environment's actual security. Landscapes are heterogeneous mosaics of interacting ecosystems, integrating land with human activities expressing aesthetic cultural values (Wiens, 1999). Mindscapes are defined by a combination of different elements, including perception, knowledge (encompassing conscientization and awareness), worldview, and values (priorities, expectations, preferences and motivations). Both landscapes and mindscapes encompass the idea of the complex set of interactive processes that take place to form them. Mindscapes are complex structures that enable perception and cognition processes, which are at the basis of our decision-making processes. These decision-making processes are not linear, and are influenced by a series of factors, all of which are important to consider. These factors can be combined within three major components: a biological, an external and an internal component (Marzall, 2006).

The mindscape features that constitute environmental resilience are characterized by the perception of security and capacity to deal with uncertainty, and can be defined as the capacity to ask 'what if...?'. They also involve the perception of the impact imposed on the ecosystem by our actions. This perception involves the capacity to ask 'what then...?' Both aspects involve external and internal components of the mindscape.

The capacity to deal with uncertainty, or to ask 'what if...?', encompasses mainly the perception of possibilities, and the ability to make choices. Raising questions permits one to go outside the box, to find a different manner in which to view reality, particularly what we consider problems or uncertainties. Being able to consider an alternative logic is an important step to widen the range of possible paths to follow, hence increasing resilience. The aspects that structure the capacity to ask 'what if...?' are mainly knowledge, creativity, and autonomy (Marzall, 2006).

Asking 'what then..?' (Orr, 1992) is mainly the awareness of the consequences that result from our actions. It considers feedback loops as guides to further action. When considering our relationship with the environment, asking 'what then..?' implies considering potential outcomes of an action, based on past experiences and acquired knowledge, as well as considering the indicators of change that arise while actively considering other possible paths. Asking 'what then...?' also involves knowledge of the environment and of possibilities, awareness of issues, awareness of consequences, responsibility, and care (Marzall, 2006).

Below we present some initiatives and the consequences they imposed on farmers' understanding of environmental impacts and on their behavior, as well as the impact these policies could have on the environment. These initiatives are Canadian Agro-Environmental Policy, Class Structure and Cross-Compliance measures.

MULTIFUNCTIONAL AGRICULTURE

The concept of landscape multifunctionality acknowledges that landscapes play multiple roles in meeting socio-cultural, production and ecological needs. Such production functions as forestry, agriculture, hunting and water consumption are included in multifunctionality. Ecological functions include support for biodiversity and natural habitats, and the recharging of groundwater. The social or information functions of recreation, cultural heritage, aesthetics and regional identity are also part of landscape multifunctionality (Brandt and Vejre, 2004; Wilton, 2005).

Wilton (2005) suggests that the opposite of multifunctionality is monofunctionality, which has been widely criticized for causing many social and environmental problems (Pimentel, 2008). The many functions of landscapes, as mentioned above, are societal, economic and natural. Ideally, these factors should be integrated, for example, recreation and tourism could be integrated with agriculture.

Along with growing urbanization, proliferating environmental interest groups and greater concern for food safety, there has been a rising interest in the multifunctionality of our agricultural land. This may generate policies that fall more in line with those of the EU where farmers are paid to produce environmental goods and services (EGS). The production of EGS has been accompanied by an expanded role for agriculture from the production of commodities to include the production of non-commodities, hence the term multifunctionality (Durand and Van Huylenbroeck (2007).

Ontario farmers have been promoting the slogan "farmers feed cities" but Sparling (2006) has pointed out that urbanites do not now feel that they are in jeopardy because of the threat that they may not have enough food. Urbanites may, however, be unaware of the extent to which farms remove carbon dioxide from the air, put in buffer strips to prevent polluted waters, protect biodiversity by providing habitat and wetlands, are beginning to harvest the wind for energy and generally provide other EGS that ameliorate the worst effects of climate change. Farmers are becoming more environmentally friendly in their production of not just food and fiber but EGSs that are valued by society (Denhartog, 2007). Though pilot projects designed to pay farmers for producing EGS in Ontario exist, the idea is still far being implemented on this side of the Atlantic.

Regarding government support for agriculture, Charlebois and Langenbacher, (2007: AA8) claim that in 2006 "federal and provincial governments spent an average of $3.53 in subsidies for every dollar earned on Canadian farms." These subsidies are still far less than in the U.S. and Europe and are tied to food production instead of rural development or the production of non-commodity EGS as in the EU. But while some authors argue against the view that there is a cheap food policy (Miller and Coble, 2005), others point to the cheap cost of energy as the source of cheap food (Lang, 2007). Certainly as oil prices fell in 2008, food prices fell as well.

The multifunctional term was first introduced as part of the United Nations program for sustainable development, Agenda 21. While the EU has made multifunctional agriculture key to its WTO arguments in favour of supporting agriculture and rural development, American negotiators have opposed the notion seeing it as a justification of unfair trading practices. "The European Commission has aimed to present *multifunctionality* as the *leitmotiv* of European rural and agricultural policy" (Gallardo *et al.* 2002: 171). Thus the US has tended to consider multifunctional agriculture as "merely an attempt to maintain the distortion of internal protection policies on the world market" (Delgado *et al.,* 2002: 28).

European proponents see multifunctional agriculture as a necessary convergence of agriculture with rural development. It is one way of reinvigorating rural society, along with other aspects of rural development including protecting watersheds, biodiversity and the environment, improving animal welfare, encouraging tourism and crafts, campsites, providing rural services such as horse stables, renovating villages while protecting rural heritage and, of course, marketing quality agricultural goods.

Belletti *et al.,* (2002: 57) observe that the implications of multifunctional agriculture are:

> quality production, new and shorter chains linking producers and consumers, organic farming, ecological sound management of local environments and landscapes by farmers, integration of care activities

into farms, involvement in new forms of energy production, agri-tourism and low input farming.

Rural tourists especially appreciate the improved rural landscapes that result from multifunctional agriculture.

The European Commission (EC) considers the most important factors comprising multifunctionality to be "1) food production, 2) the defense of landscape values and those of the rural environment, and 3) the contribution to the viability of rural areas and to a balanced economic development from a territorial viewpoint" (Gallardo *et al.* 2002: 172).

Even if the enterprise is non-competitive, it may be viable, excluding the value of the land, if it is able to maintain its productive capital. This could be achieved by selling not only its primary produce but its green services such that the natural resources are preserved and the land's future as a source of livelihood is safeguarded. Nevertheless, as Swagemakers (2002: 191) points out, "combining new functions with food production demands a reorganization of farming practices."

Despite the extent of soybeans, corn and livestock, because southern Ontario has most of Canada's agro-climatic ratings, as well as excellent young soils, the region supports the greatest diversity of crops in Canada. Recent provincial developments attempting to protect green space and farmland in the face of growing urbanization have impacted southern Ontario farmers significantly. These provincial government initiatives include the development and protection of a Greenbelt, greater efforts at urban Growth Management ('Places to Grow') as well as Planning Act changes. These efforts are part of the growing move to regulate growth, industry and agriculture in the face of predictions that the Greater Toronto Area will increase to 10.5 million by the year 2031 (Wilton, 2006).

POLY-OCULAR ADVOCACY FOR STRONG SUSTAINABILITY

Since the authors of this book are the observers, our starting points-of-view cannot be other than the human one, which is therefore anthropocentric (von Foerster, interviewed by Poerksen, 2003), regardless of how much we may wish to transcend our anthropocentrism. This does not preclude other existing viewpoints, but suggests that the existence of every individual as a separate starting point-of-view or observation[1]. We also recognize that there are limits to our perspectives (Morito, 2002) but our analysis presents the phenomena descriptively and relationally given the context to the best of our ability. We also seek to be open to other perspectives because they can bring forth understandings that we were unaware of earlier and this is as a result of our poly-ocular position (Maruyama, 1978, 1985, 2004). Conflict is a natural result of the diversity of perception processes and starting points-of-view (Sharov, 1992; Robert, 2000). Thus our perspectives may not only differ from each other within this book but they may also conflict with others including the differing perspectives of farmers of different sized operations and commodities, farm managers and workers, government personnel and other academics, yet diverse opinions are pivotal for the possibility for creative insights and good communication of knowledge (Pagels, 2003). The challenge is to find shared perspectives and distinctions defining some common ground to act while maintaining our uniqueness.

Acknowledging an anthropocentric point-of-view emphasizes first that this book is an outcome of our observations and that as much as we consider the environment, we see it from our own personal perspectives. While we try to consider farmers' view-points and to take a 'neutral' standpoint during observations, the information we gather will always be filtered through our own perspectives (Marzall, 2006). Regardless of our attempt to understand farmers, we are academics and, to a lesser extent, environmental officials, not farmers. This limits our ability to grasp what farmers must do to survive economically as well as to preserve their quality of life and their environment.

One of the problems that must be acknowledged when we realize that we find it difficult not be anthropocentric to some degree is the superior position anthropocentrism accords humans in the biosphere. Ecocentrism (also known as biocentric egalitarianism) presents an alternative perspective by pointing to the intrinsic value of all forms of life. Ecocentric positions do not privilege people over any other species and this view is a form of Ideal Sustainable Development. Nevertheless, most of our present resource and environmental policy remains substantially anthropocentric. The belief that humans should remain central to the control of the environment and resources is a form of utilitarian anthropocentrism. By contrast, our position is closer to the desire for strong sustainable development which recognizes the importance of protecting biodiversity as well as the need to seek a balanced approach to the social,

---

[1] Neither do the editor and all of the authors in this book agree on everything in this book.

economic and environmental components of intensive agriculture in the face of urban sprawl (Hessing *et al.*, 2005).

AGRO-ENVIRONMENTAL POLICY

Environmental policy is policy relating to consumption, production and resource extraction while resource policy refers to regulations governing primary resource extraction. Government environmental and resource policy affects everyone whether or not they own private property because clean air, water and land are essential to our health and well-being. Water, air and land resources require management yet Hessing *et al.* (2005) have shown that Canadian resource and environmental policy from before 1800 to the present moved from inaction, to a concern with extracting resource rents or fees, to resource conservation and finally to resource management. At first, producers were asked to regulate themselves, which is essentially what most Ontario farm organizations continue to insist upon, standing behind the Environmental Farm Plan as their guarantee of environmental security. Self-regulation of industry was however, superseded by greater state environmental regulation of industry, which has recently spread to Ontario agriculture with legislation like the 2002 Nutrient Management Act (NMA), 2005 Greenbelt Act and the 2006 Clean Water Act (CWA).

In acting as it did to limit the requirement of a nutrient management plan (NMP) for only the intensive livestock operations (ILOs), the Ontario Farm Environmental Coalition was behaving like most other resource industries that have historically resisted most regulation falling back on private sector solutions to problems that affect common as well as private property resources. Of course, ILOs ought to be regulated as they are large operations that normally store their animals' liquid manure in lagoons until it is time to spread the manure onto fields for use as fertilizer. If there is a problem with their lagoons' construction including over-application during the spreading of the manure, local creeks may be polluted with excess nitrates and phosphates as well as bacteria. If these creeks are the headwaters of major rivers like the Grand, Maitland and Ausable and Bayfield rivers, the harmful effects of the excess nutrients will travel into the Great Lakes and damage fisheries along the way. But the same can be said for smaller operations including cash crop farms because a large number of them collectively can also produce excess nutrients and chemical runoff. Until now, however, they have escaped without having to develop nutrient management plans.

Government pressures to deregulate achieved a peak in Ontario during the Harris Government in the late 1990s when, for example, the Ministry of the Environment was diminished by a third (Montpetit, 2003). This anti-regulationist approach ended with the shocking Walkerton disaster of 2000 and other related problems. The Liberal Government that followed has been re-regulating in an attempt to implement Justice O'Connor's recommendations. As well, improved technology has been enhancing our ability to reduce harmful effluents. Serguei Golovan and others at the University of Guelph, for example, have developed the "low phosphorus pig" though a patent for promoting the spread of this animal has yet to be obtained.

At least since Rachel Carson's groundbreaking critique of the effects of chemicals and pesticides on biodiversity in *Silent Spring* there has been a growth in environmental concern among the public (Inglehart, 1977, 1990). This has tended to challenge industrial claims that we should simply allow the market to determine what happens with resource industries. However, as Hessing *et al.* (2005) have observed, despite the growing environmental consciousness among the Canadian public and the spread of environmental interest groups, the public plays a limited role in setting the environmental agenda relative to the State and productive, in this case, agri-food forces.

Increasingly, however, the Canadian State has tried to move to a more proactive environmental agenda shifting the policy agenda to a more precautionary principled preventative approach. Providing government funding for Environmental Management Systems (EMSs) is one way of implementing the precautionary principle. Provincially, the NMA embodies the precautionary principle that characterizes the Environmental Farm Plan (EFP) though the NMA is regulatory and the EFP voluntary. However, voluntary environmental risk management behavior has often been less effective than government regulation because many producers choose not to participate and non-compliance is usually not punished. Within agriculture, significant government/producer conflict remains, including internal government conflict regarding mandatory versus voluntary environmental regulation. One obvious problem with the regulatory approach is that it tends to be top-down and therefore generally non-participatory. Another downside of regulation is that it can also lead to a 'one size fits all' type of regulation, ignoring important differences among producer operations. Of course, if government had more agricultural extension agents they could involve farmers more in understanding the meaning of the regulations but the Harris

Government of the mid-1990s managed to gut the majority of the remaining publicly funded agricultural extensionists then still in existence.

An alternative to the undervaluation of externalized costs has been developed by ecological economists who would like to see externalities from production, like soil erosion and nitrate leaching, included as part of the costs of production. They would also like to see an anthropocentric environmentalism tempered by biocentrism. Such a perspective would emphasize the importance of maintaining biodiversity emphasizing aesthetic values and focusing on protecting the environment in general (Hessing *et al.*, 2005).

Public environmental policy is heavily influenced by particular issues like the Walkerton tragedy, which have captured public attention and govern action (producing the Nutrient Management Act (NMA) and the Clean Water Act (CWA) among other regulatory structures). Public opinion on the other hand, has only an indirect effect on policy though the growing environmental movement, periodic disasters and occasional sensationalist reporting typifies the media's environmental coverage. Nonetheless, public interest groups like the Sierra Club and Pollution Probe tend to react to environmental problems like the Walkerton tragedy, which can help to galvanize the Government into action as illustrated with the NMA and CWA (Hessing *et al.*, 2005).

Because Canadian farm policy depends entirely on providing assistance to farmers in a manner which is "linked to the production of specific commodities" this tends to encourage farmers to continue to produce those commodities and it rewards farmers in relation to their scale of operation (Charlebois and Langenbacher, 2007). Thus small operation farmers who also produce EGS are not receiving much support through CAIS. There is also still not enough support for marginal farmers to make the transition out of farming into other employment. We return to this in chapters 3 and 7.

SUMMARY

This chapter considers the nature of agro-environmental security, the implications this imposes on environmental regulation and how farmers' voluntary environmental behavior influenced by their mindscapes define their interaction with the environment, impacting the sustainability of farming systems and the integrity of ecosystems. In Canada, some would like to be able to brand ourselves as producing food in an ecologically friendly manner (e.g. Deloitte, 2007), and are encouraged to do so by our increasingly organically oriented urban consumers. Both federal and provincial governments in Canada have made strides in this direction, while simultaneously greater efforts to protect environmental security are being required of agriculture but at present our system falls short in many important policy and behavioral areas. There are also now many good reasons to reform our existing farm support programs and provide better ways of supporting the positive impacts of agriculture on the environment.

The latest international food regime prioritizes a few commodities like corn, soybeans and livestock processed and peddled to consumers, often as calorie-dense fast foods with insufficient nutrient value which encourage the development of obesity, diabetes, arthritis and hypertension. Despite being an increasingly multicultural society, including for example a growing vegetarian South Asian population, many nutritious fruits and vegetables that could be grown locally are not. The contradiction between the dominant international food regime and human health may be accelerating changes which spawn more organic food production and the local food movement with community shared agriculture, farmers' markets all of which may culminate in a lower carbon footprint and healthier ecosystems.

In undertaking this research we acknowledge that we have a somewhat anthropocentric environmental orientation but one which favours strong sustainability within our agro-ecosystems. This approach acknowledges that landscapes are perceived human concepts embodying people as well as biodiverse ecosystems. Understanding how people's mindscapes interact with their landscapes and ecosystems helps us plan for more sustainable present and future change. Environmental concern is associated not only with education and income but also to some extent with particular mindscapes which in turn are found in different proportions depending on social class and culture.

As researchers we view agro-ecosystems neither from a purely ecocentric nor anthropocentric point of view but from one that acknowledges that humans must do a better job to prevent climate change and protect biodiversity while continuing to produce food and fiber efficiently and effectively. We are concerned about the lack of debate about the extent to which we are controlled by the latest food regime including the types of foods that have come to dominate the supermarkets and fast food chains.

We also recognize that there are diverse social class, ethnic and gender perspectives which impact cultural landscapes. Agro-ecosystems are central to these complex systems' resilience. Environmental resilience is in turn a function of how farmers' mindscapes impact the social and environmental resilience of these agro-ecosystems. The balance between voluntary and regulatory management, the multifunctionality of landscapes and the possibility, discussed in chapter three below, that the rise of cross-compliant systems of management in Europe and Quebec are all issues that may soon have a more significant effect on Ontario's approach to agroecosystem management.

In: Agricultural and Environmental Security...   ISBN: 978-1-61668-156-2
Editor: Glen Filson   ©2011 Nova Science Publishers, Inc.

Chapter 3

ONTARIO WATER QUALITY LEGISLATION, PROGRAMMING AND ALTERNATIVES ABROAD

Pamela Lamba, Paige Agnew, Glen Filson and Bamidele Adekunle

INTRODUCTION

This chapter considers the environmental acts, statutes, agricultural policies and programs that have developed in Canada and Ontario over the past couple of decades. Water quality assurance programs and nutrient management legislation are reviewed with comparative references to North American and European programs. This provides essential background for understanding the incentives and pressures affecting farmers' adoption of environmentally beneficial management practices which are the focus of the case studies which follow in chapters 4-6. As will be seen, recent threats to our water and food supplies have had the effect of accelerating the construction of environmental regulations which affect farmers and farmers have to some extent mobilized against these regulations as a result.

ENVIRONMENTAL LEGISLATION

After examining some highlights of recent environmental legislation in Table 3.1 we then review some of the most important recent environmental programs affecting farmers.

Table 3.1 Ontario Environmental Legislation Since 1990

Legislation	Description	Date
The Ontario Water Resources Act	**Protect quality and quantity of Ontario's surface and ground water resources** Administered by the MOE Standardizes sewage disposal; forbids discharge of materials that may impair water and regulates "water taking" Most recent amendment: O. Reg.60/08 – Lake Simcoe Protection – amended by O. Reg. 130/09	1990
Public Lands Act	**Administered by MNR** **Regulates the use, management, sale and outlook of public lands and forests**	1990
Environmental Assessment Act	**Assesses major public or designated private projects** Primary environmental planning statute Issued my the MOE **Evaluate ecological, social, cultural and economic impacts**	1990, then 2006
The Environmentl Bill of Rights	**Protect, conserve and restore environment** Sustainability of the environment Protect the right to a healthful environment Ensure public participation in environmental decision-making **Increase citizen's access to the courts for environmental protection**	1993
Nutrient Management Act	**Facilitates the regulation for nutrient management through farm plans and strategies** Administered by the MOE and OMAFRA Control nutrients on farm from entering into surface or ground water Disposal of Dead Farm Animals	2002, revised in 2003 and 2009
The Safe Drinking	**Relates to the treatment and distribution of drinking water**	2002

Water Act	Protects human health through the control and regulation of drinking-water systems and drinking water testing Legally-binding standards for contaminations; mandatory to use licensed laboratories for testing the water and mandatory to report adverse test results Administered by the MOE	
Greenbelt Act	**Protecting green space and containing urban sprawl** Administered by MAH Improve and enhance quality of life for both rural and urban residents **Identifies where urbanization should not occur to provide permanent protection to agricultural land base and ecological features**	**2005**
Clean Water Act	**Multi-barrier approach to provide a safe supply of drinking water Phase 1: Evaluate the potential sources of contamina- tion, the quantity of water available and vulnerable bodies of water Phase 2: Creating programs, policies and initiatives to minimize the risk of contamination and to ensure long- term protection**	**2006**
Endangered Species Act	**Identifies and protects species and their habitat which are at risk partly through the levying of significant fines against transgressors.**	**1990, replaced in 2007**
Pesticides Act	**Cosmetic pesticides are banned for non-agricultural use** Exceptions also exist for golf courses, specialty turf, specified sports fields and arboriculture (http://www.ene.gov.on.ca/en/land/pesticides/factsheet s/fs-agriculture.pdf)	**Apr, 2009**

In 1993 the New Democratic Party (NDP) Government passed the Environmental Bill of Rights (EBR) which was designed to (a) protect, conserve and restore the integrity of the environment, (b) provide for the sustainability of the environment, and (c) protect the right to a healthful environment. They also passed the Farm Registration and Farm Organizations Funding Act which required all farmers to belong to either the Ontario Federation of Agriculture or the smaller Christian Farmers Federation of Ontario. The effect of the EBR was to galvanize farm organizations into forming the 38 farm organization member Ontario Farm Environmental Coalition which shortly thereafter prepared a document called "Our Farm Environmental Agenda" in early 1992 (Monpetit, 2003). Later that year, these groups then developed the Environmental Farm Plan (EFP) which they adapted from Wisconsin's Farm*A*Syst program (see below). Jeff Rose, an early advocate of the EFP used to sell it to farmers by saying "Who would you rather have in your face? Other farmers like me or the Government? (Rose, 1995). The NDP was swept out of office, especially by rural ridings, in 1995 mainly in reaction against their deficit spending designed to lesson the effects of the 1991 recession. The new Conservative Government passed the Farming and Food Production Protection Act which gives farmers "protection against nuisance lawsuits over dust, noise, and odours as long as they adhere to "normal" farm practices (Montpetit, 2003). While this Act protects farmers, it directly contradicts the spirit of the Environmental Bill of Rights (EBR). One lasting positive result of the EBR, however, was the creation of an Environmental Commissioner to monitor the Province's compliance with the protection of the environment. The Office is held by Gord Miller and his yearly reports have been very effective in giving environmental direction to the Government. He has also been a major supporter of the Nutrient Management Act (2003) and the Clean Water Act (2006) passed by the Liberals, who in turn replaced the Tories who had been tarred with the consequences of their radical downsizing of the Ministry of the Environment as well as the Walkerton tragedy among other efforts to slay the fiscal crisis of the state via cutbacks.While the period of self-regulation of farmers' environmental behavior had continued throughout the Conservative Government's period in office, the situation changed dramatically after the Walkerton tragedy. Whereas the Ontario Ministry of Agriculture, Food and Rural Affairs (OMAFRA), the Ministry of the Environment (MOE) and the Ontario Ministry of Municipal Affairs and Housing (MAH) had administered an Agriculture Code of Practice this code was merely a set of guidelines for livestock operations to minimize land, water and air pollution potential. The way it had worked was: if requested by the landowner, an application for a Certificate of Compliance could be made, and an inspection could of the farming operation would follow. If the conditions outlined in the Code of Practice were fulfilled, then a Certificate of Compliance would be issued

(OMAF, 2005). The acceptance of Justice O'Connor's recommendations by the new provincial Liberal Government meant that farmers were more than simply subjected to voluntary guidelines for "managing nutrients and minimum distance separations" (Montpetit, 2003: 94). In 2002, the Safe Drinking Water Act was introduced following Justice O'Connor's *Report of the Walkerton Inquiry* to deal with the matters relating to treatment and distribution of drinking water. The main purpose of the Act is to protect human health through the control and regulation of drinking-water systems and drinking water testing. The Act generally establishes legally-binding standards for contaminants in drinking water; makes it mandatory to use licensed and accredited laboratories for drinking water testing; requires the reporting of adverse test results and for taking "corrective action". It also requires that drinking water system operators be trained and certified and be administered by Ministry of the Environment (MOE, 2004).Bonnet *et al.* (2005) complained that, after the Walkerton tragedy, and despite the farmers' earlier cooperative approach to environmental management, the then Liberal Government unilaterally adopted an approach that involved putting "hard and fast" regulations in place. Farmers were excluded from the initial approach of establishing these regulations and they objected to the draft regulations that were announced. Eventually the NMA was revised in 2005, restricting its immediate application to farms with over 300 animal units plus new and expanding livestock farms. All local municipal by-laws are over-ridden by the Act so some are amending their original nutrient management by-laws though the majority have still not changed them (Caldwell, Mar. 9, 2005). The changes that were made to the NMA aligned it better with the CWA though more changes may be needed after the final version of the CWA was passed in 2006. The CWA has been written "so that Drinking Water Source Protection legislation will supersede Nutrient Management legislation" (Bonnett *et al.,* 2005: 68). The revised NMA is now more risk oriented so that the greatest response will be linked to the areas where the risks are highest and funding will be directed to where the greatest possible risk reduction will occur[1]. The Source Protection Planning provided for in the CWA takes a collaborative, watershed approach to risk based priority setting by relevant stakeholders.

The Nutrient Management Act (NMA) (2003) provides a framework for implementing standards for nutrient management on farms. It can be used as a

---

[1]From the Federal Government there is cost sharing up to $30,000 split on a 30 to 50% basis. Cost-sharing between the federal and provincial governments has provided up to 90% of funding for manure storages so farmers can get up to $60,000 for manure storage and handling.

tool to balance crop needs with the additives coming from livestock or other sources. The Act is designed to facilitate the regulation for nutrient management through the creation of farm plans and strategies, and is administered jointly by the MOE and OMAFRA. It is intended to control nutrients on farms from entering into surface water or infiltrating into groundwater such as through setback distances and construction standards for storage. It is also designed to control pollution from biosolids, such as sludge from sewage treatment plants when they are spread on land (OMAF, 2005). Regulatory approaches, like the Agricultural Policy Framework nationally, and the Nutrient Management Act and Clean Water Act, demonstrate the federal government and the province's desire to maintain higher environmental standards than existed earlier. The Walkerton tragedy had awakened the public to the environmental and health related dangers of "business as usual" despite increasing intensification of agriculture (Filson, 2004b).

In 2005, the Province passed the Greenbelt Act which provides zoning regulation to protect greenspace and contain urban sprawl. The Act's main goal is to improve and enhance the quality of life for both urban and rural residents of Ontario. It allocates a broad band of countryside or "greenbelt" that will be permanently protected under this legislation/plan. The Act's main objective is to provide for a diverse range of rural communities for agriculture, tourism, recreation and resources uses, while also protecting the ecological integrity of the natural systems present. In other words, the Greenbelt Act identifies where urbanization should not occur in order to provide permanent protection to the agricultural land base and the ecological features and functions occurring on the landscape (MMAH, 2004).

While the Greenbelt Act is very popular with urban Ontario residents wanting to protect green areas and biodiversity, many farmers see the Greenbelt as an almost exclusively urban idea. Zoning can obviously have a negative impact on the price of land, which is among farms' most important asset and some farmers have experienced a reduction in the value of their land by as much as one-third so many farmers and farm organizations have become very concerned about the Act and its possible expansion (Filson and Sarker, 2008). A related piece of legislation is the Places to Grow Act enables the government to plan for population growth, economic expansion and the protection of the environment, agricultural lands and other valuable resources in a co-coordinated and strategic way (*Places to Grow*, 2005).

CURRENT AGRI-ENVIRONMENTAL PROGRAMS IN ONTARIO

The most significant agri-environmental programs over the past couple of decades are presented in Table 3.2.

Table 3.2 Past Agri-Environmental Programs

Program	Description	Date
Soil and Water Environmental Enhancement Program (SWEEP)	Canadian and American governments involved Control phosphorus pollution entering the Great Lakes from industrial and agricultural Sources	1986-1993
Land Stewardship Program (LS)	Two parts: (LS I & LS II) Provides grants to farmers who have adopted conservation farming practices A grant of 2.25 million first 3 years for maintenance, training and equipment; $38 million granted for farmers in the last 3 years	1987-1990 & 1990-1994
Great Lakes Water Quality Program (GLWQ)	Canada and American government involvement Focused on agricultural chemicals and their reaction within the environment Included water quality issues and conservation practices	1989-1994
The National Soil Conservation Program (NSCP)	Assist in the implementation of on-farm conservation measures; provide research and monitoring and to increase awareness of the benefits of these programs	1991-1993
Land Management Assistance Program (LMAP)	Federally funded-$15.24 million Extension of the NSCP Funds for Rural Conservation Club Programs and development of BMPs Booklet Helped in the pilot project for the EFPs and the Workbook/Video Package	1992-1994
Environmental Sustainability Initiative (ESI)	Adoption of sustainable farming practices as a means to increasing long-term financial returns to producers. Sustain agriculturalists' natural resources foundation and competitiveness First to initiate the EFP Pilot project	1992-1993
Green Plan	Federal program that used community consultation to identify issues and develop methods of action 30% of funds went to research; 45% to farmers and their organizations and 25% to technology transfer	1992-1997

Table 3.2 Continued		
The Clean Up Rural Beaches Program	Provide financial and technical assistance to rural landowners who voluntarily implement BMPs Control non-point source pollution Increase public awareness of agricultural activities on water quality Administered by the MOE	1991-1996
Healthy Futures for Ontario Agriculture	The program is designed to encourage partnerships in Ontario's agriculture and food sectors in order to: (1) improve the quality and safety of Ontario food products through the adoption of new or upgraded technology and/or BMPs; (2) capitalize on marketing and export opportunities and (3) safeguard rural water quantity and quality	2000-2004
Rural Water Quality Program	Administered by the GRCA Offers financial assistance to landowners to share the cost of implementing BMPs that improve water quality and provides technical assistances Helps to respond to particular water quality issues on their farms Voluntary program	1998-Now
Environmental Farm Plan	Voluntary documents prepared by farmers to increase the knowledge of the environment on their farms Identify environmental strengths; environmental concerns and place goals and timetables to improve environmental conditions Led by OFA, CFFO, OFAC and AGCare	1993 – Now
Agricultural Policy Framework I	Composed of 5 elements: business risk management; food safety and food quality; science and innovation and environment and renewal qualities All provinces have signed the APF Sector-wide consolidation to eliminate redundancy & duplication and establish universal standards of environmental quality Continue sustainability with the environment by improving environmental planning tools and management systems Administered by AAFC	2001-2007
Agriculture Code of Practice	Administered by OMAF, MOE and MAH Provides guidelines for livestock operations to minimize land, water and air pollution potential Farms are inspected and if conditions are fulfilled then a Certificate of Compliance is issued	2001-now

These programs have usually provided a combination of incentives and education to help farmers adopt more environmentally friendly behavior.

For instance, the Clean Up Rural Beaches (CURB) program was developed when a number of Ontario beaches in the early 1980s were periodically closed for swimming due to high concentrations of fecal bacteria from livestock. This bacteria that was found in watercourses and beaches was linked to livestock's access to watercourses, runoff, poor manure spreading practices and illegal disposal of milk-house wash water. In addition to this, rural septic systems were also a potential source of the contamination (and to some extent still are). CURB's main goal was to provide financial and technical assistance to rural landowners who voluntarily implemented Best Management Practices (BMPs). The program was developed to control non point source water pollution in 17 watersheds in Ontario. The program provided a $50 million dollar incentive to landowners for implementing water quality projects (Ryan, 1999).

Some programs have involved Canada and the United States in collaborative efforts to clean up the Great Lakes by, for example, reducing the amount of excess nutrients from farming. The Soil and Water Environmental Enhancement Program (SWEEP) was developed and negotiated by the Canadian and American governments. The main purpose was to control phosphorous pollution entering the Great Lakes from industrial and agricultural sources (OMAFRA and AAFC, 2002). An evaluation of the project indicated that farmers had a good stewardship ethic and were willing, if they had not done so already, to implement conservation measures on their farms. However, a major limitation with this program was the financial constraints that were found to be a barrier, as well as how the changes required would affect farmers' practices (OMAFRA and AAFC, 2002).

The Land Stewardship Program was separate from SWEEP, but it also gives grants to farmers to encourage them to adopt conservation farming practices. The main objective of the program was to improve soil resources and water management by reducing soil erosion and compaction, restoring soil organic matter and tilth, and reducing the potential of environmental contamination arising from agricultural practices (OMAFRA and AAFC, 2002). For the first few years a grant of $2.25 million was available for soil building and maintenance projects, structures, technical training, machinery and equipment. The second phase of the program included an additional $38 million grant which was allocated to farmers.

The National Soil Conservation Program was developed in 1991, and was meant to assist in the implementation of on-farm conservation measures;

provide research and monitoring and increase awareness of the benefits of these programs. The Federal and Provincial governments were both involved, along with consultancies and farm organizations, and were coordinated by the Prairie Farm Rehabilitation Act (OMAFRA and AAFC, 2002). In the same year, another program was introduced, Land Management Assistance Program (LMAP). This was a federally funded program ($15.242 million of funding). The program was developed to encourage farmers to implement environmentally sound management practices on their land (OMAFRA and AAFC, 2002). The LMAP was an extension of the NSCP. It was through this funding that the pilot project for the Environmental Farm Plans (EFPs) and the development of the Workbook/Video package was created at the University of Guelph in consultation with farmers.

Both the Federal and Provincial Government have focused on promoting sustainable agricultural practices. The program was one of the first to develop Best Management Practices Series booklets (OMAFRA and AAFC, 2002). This trend continued through the Land Management Assistance Program, which was given $15.242 million over a two-year period from 1992 to 1994. The goal was to promote the adoption of sustainable farming practices as a means of increasing the long-term financial returns to producers [crops]; the sustainability of agricultures' natural resources foundation, and the maintenance of Ontario agri-food sector competitiveness (OMAFRA and AAFC, 2002).

A major strength of the Environmental Farm Plan (EFP) program mentioned above as having begun in 1993 is that it works with individual farmers to identify environmental risk on their farms and then develop solutions that will provide maximum protection for the environment. In April, 2005 the AFFC announced a $57 million funding for EFP, to enable farmers in Canada to protect the environment while conducting their farming operations (OFA, 2005). Instead of providing a mere $1500 to each farmer doing an EFP as in the past the new program provides up to $30,000 for individual farmers to take corrective steps on their farms. The Canada-Ontario EFP permits farmers to contribute their own labour and equipment as part of the agreement, which has helped farmers gain confidence in how the new program is administered. The OSCIA continues to deliver the new program on behalf of the Ontario Farm Environmental Coalition, which includes OFA, the Christian Farmers Federation of Ontario (CFFO), the Ontario Farm Animal Council and AGCare (Agricultural Groups Concerned About Resources and the Environment) (OFA, 2005) though funding has been cut back somewhat since then and farmers lack confidence that EFP funding will remain in place.

The AAFC's National Environmental Farm Planning (NEFP) initiative seeks to have every farm in Canada do an environmental farm scan. There are two types of initiatives identified under the NEFP: individual EFPs and equivalent agri-environmental planning (EAEP). An EAEP is similar to "individual farm planning, but is implemented by an organized group of producers on a multi-farm basis determined either by commodity or by geographic area, such as a watershed or other ecological zone" (Canada's Agricultural Policy Framework, 2005).

As a result of better appreciation of environmental sustainability, the Healthy Futures for Ontario Agriculture Program's was introduced to maintain and build on the success of Ontario's agri-food industry. The program was a $90 million, four-year project that contributed annually $25 million to the provincial economy, generated $6.2 billion in agri-food exports and employed more than 640,000 people. The program was produced by the Ministry of Agriculture and Food in co-operation with people involved in the province's agri-food industry. Unfortunately, incentives for the program ended in 2004 (OMAF, 2005).

Other programs at the municipal and conservation authority level do continue, however, such as the extremely successful Rural Water Quality Program (RWQP) launched by the Grand River Conservation Authority in 1995 which has since spread to other municipalities and conservation authorities. The RWQP offers financial assistance to qualified landowners to share the cost of implementing best management practices that help improve water quality (GRCA, 2004).

The RWQP was formed through input from steering committees that included local farmers, the Ontario Farm Environmental Coalition, the Ontario Federation of Agriculture, the (Ontario Soil and Crop Improvement Association (OSCIA), local farm groups, OMAFRA, MNR, MOE and local municipalities beginning with the Waterloo Region (Ryan, 2006). Besides Waterloo, Brant County, Brantford, Guelph and Wellington County each contribute $750,000 to the RWQP. Once the farmer has completed an EFP, and the Local Peer Review Committee has approved it, then the EFP will help farmers assess the environmental strengths of their farm, identify areas of environmental concern and set practical goals to improve conditions in relation to the farmers' timetable. There are higher incentives for things without an obvious agricultural benefit so farmers can receive 100% of the cost of fencing their cattle from a stream. More than 1,600 projects had been completed since 1998 (Ryan, 2007).

The Clean Water Program (CWP) is a recent form of technical and financial assistance available to farmers in southwestern Ontario in order to help them improve and protect water quality. It is funded by local municipalities, and delivered by local Conservation Authorities. The CAs involved with the CWP include, at the least, Ausable Bayfield, Catfish Creek, Grand River, Kettle Creek, Long Point Region, Lower Thames Valley, Maitland Valley, St. Clair Region, the Upper Thames River CA and the South Nation CA. Like the GRCA's RWQP the CWP is linked to participation in the EFP. If the project is in a Wellhead Protection Area as specified by the CWA, or if the project also qualifies for EFP funding, the farmer is eligible to receive up to 70% of their funding through the CWP. Projects funded by CWP include milkhouse washwater disposal, clean water diversion, livestock access restriction to watercourses, fertilizer, chemical and fuel storage or handling, NMPs, wellhead protection, septic systems, erosion control structures and fragile land retirement (Clean Water Program, 2007). Even funded on the usual 50%-50% basis, some of these projects qualify for up to $5,000 in municipal funding.

How do these agri-environmental programs and environmental legislation compare with those in Europe and the United States? The following sections reveal what we learned about what they have been doing to protect agri-environmental security.

ENVIRONMENTAL CONCERNS ABROAD EUROPE

The European Union's Common Agricultural Policy (CAP) was developed originally to prevent food shortages and to ensure farmers were self-sufficient (European Union, 2004). Throughout the years, different legislation was implemented to reduce the amount of subsidies and ensure that more direct payments were being made to farmers. Subsequently, the implementation of Agenda 2000, ensured greater environmental and rural development focus, by focusing on the health of all Europeans (European Union, 2004).

Many EU countries have implemented individual plans, which were in existence before the umbrella legislation. They are also not forced to adopt the environmental policy of the CAP, regulation 2078, as part of their own environmental policies (Buller et al, 2000). The CAP is an agricultural policy which provides farmers with a reasonable standard of living, consumers with quality food at fair prices while at the same time preserving the cultural heritage of the society (Wikipedia, 2008). CAP characteristics include the use of import tariffs, which imposes a tariff on agricultural products from non-member countries. There is also an import quota, where countries will have to negotiate the quantity they will import to the EU. Production quotas and set aside payment are also administered to crops like milk, grain and wine. Since CAP is a program that protects farmers, an internal intervention price is set to stabilise the price of products and direct subsidies are paid to farmers for planting certain crops. Unfortunately these subsidies encourage farmers to plant crops even when they would be better off or more efficient in the production of other crops. As a result of complaints from different quarters, a reformed CAP was introduced in 2005, which stipulates the phased transfer of subsidy to land stewardship rather than specific crop production (2005 – 2012). Subsidies will now be based on adoption of environmentally beneficial farming methods. This new direction will give farmers the opportunity to produce crops that are in demand and it will serve as a disincentive to overproduce.

Another significant improvement to the CAP is decoupling.To decouple is to separate, disconnect or dissociate the administration of subsidies from particular crops. Subsidies are now attached to BMPs and a program called Single Payment Scheme (SPS) has been introduced which gives payment to farmers based on cross-compliance. Cross-compliance requires farmers to comply with environmental, food safety and welfare standards. The implication of this is that farmers will be better stewards of their environment by producing ecological goods and services (EGS) while at the same time producing crops and livestock necessary for the increase in their income. Decoupling and cross-compliance leads to sustainable development because the ecosystem is not affected while the farmers produce to keep up with the expectation of the ever demanding and fast moving world we live in.

Some people believe that CAP leads to unfair practice, because it supports the inefficient farmers at the expense of poor farmers in developing countries. Before the reform CAP supported big farms at the expense of small farmers because of tied subsidy and price intervention. This is now corrected because of (a) decoupling and (b) stressing the adoption of environmental beneficial farming methods. The reform has improved the well-being of small farmers.

The issue of environmental problems as a result of production based subsidy has been solved because is now the centre of farming policy. The criticism of CAP that it involves state intervention will never be solved, because it is based on the perception of free market advocates who disagree with any type of government intervention. To the free market advocates, the free market allocates resources efficiently while state intervention leads to inefficiency, although the term efficiency is also very difficult to define.

Some also say that CAP is unsustainable because of the huge amount of the EU budget which is spent on the program. Only 5% of Europeans work on EU farms and the sector is responsible for only 3% of EU's GDP (Wikipedia, 2008). The argument against CAP is whether it is reasonable to spend a lot of money a project where only very few benefit. The good thing about the reformed CAP is the introduction of payment for BMPs because everybody benefits from healthy environment and the consumption of high quality food.

The new CAP system is intended to preserve farm land, maintaining it in excellent agricultural and ecological condition by avoiding erosion and preserving the organic content of the soil while also maintaining the landscape. However, European farmers have resisted the increased regulation of their farming behavior as farmers must now adhere to basic environmental requirements dictated by the EU in order to receive direct payments from the EU. These regulations, for instance, set out rules for nitrogen and phosphorus application in return for which the farmers receive funding per hectare. The EU's requirements include environmental conditions, food safety, and health requirements for humans, animals and plants. There is also a ban on the application of nitrogen fertilizer between November 1st and January 31st (e.g. no manure on snow) and a maximum allowance of 80 kg of nitrogen per hectare or 40 kg of ammonia per hectare. Agricultural extension services are heavily involved in advising farmers on these new cross-compliance rules and therefore, these changes have required quite an adjustment for them as well as the farmers (Hessenhuber, 2006).

If the above rules are broken and this is discovered through the environmental audits conducted randomly by government environmental inspectors, farmers may be asked to reduce their cattle stock, rent additional fields to distribute their excess manure, dispense liquid manure on operations without cattle or use the excess for additional biogas production. A combination of environmental inspectors and agricultural extensionists work with farmers to help them understand how to comply with the newly implemented regulations and also how to prevent the loss of their environmental payments (Hessenhuber, 2006).

In England farmers claiming environmental payments from government due to the CAP "…are responsible for understanding and meeting cross compliance requirements" (Cross Compliance, 2006). In support of this system they can access technical advice from agricultural contractors, other land managers and various advisors. "Farmers in receipt of the new Single Farm Payment are obliged to comply with a number of pre-existing environmental, animal health and welfare directives and regulations" (A Vision for the Common Agricultural Policy, 2005: 33). Farmers are therefore required to maintain their farms in good agricultural and environmental condition in order to receive their payments. Various agri-environmental schemes including Higher Level Stewardship are available to reduce farmers' cost of compliance.

UNITED STATES

According to Risse and Tanner (2004-2005), many voluntary environmental programs can work. Payments for environmental services began first in the European Union, and have since been adopted all over the world, and many of these programs have been adopted in the United States, at the local, regional and state level. The way it works is that landowners/farmers receive a particular payment for adopting practices that enhance or protect the local environment. This approach is similar to Ontario's recent Healthy Futures Program sponsored by OMAFRA and MOE in Ontario (Agnew, 2005).

The U.S. Rural Clean Water Program is a non-point source pollution control program, federally sponsored to address agricultural non-point source pollution. It included twenty-one projects that were conducted during a 10-year period over a wide range of pollution problems and impaired water uses. The study area included 22 states and involved the implementation of BMPs to reduce non-point source pollution and subsequent water quality monitoring to measure improvement. The landowners voluntarily participated, and were offered cost sharing and technical assistance as incentives for implementing recommended BMPs (Risse and Tanner, 2004-2005).

During 1997, New York developed the Watershed Memorandum Agreement, which provided a legal framework and funding for programs that provided economic incentives for farmers who practice environmental stewardship by marinating and enhancing local water quality in the Catskills. The government had selected agriculture to be the ideal land use for its enhanced buffering and filtering capabilities for surface water infiltration as compared to urban environments, within the Catskills. There are numerous programs and initiatives that were developed to maintain the quality and quantity of New York's City's drinking water supply, and with these combined it became New York City's Watershed Management Plan (Agnew, 2005).

New York City's Watershed Management Plan includes three major programs to protect and maintain agricultural land, and to standardize and improve the environmental and agricultural practices of local farmers, in an effort to protect local waters sources. The first program is the *Watershed Agricultural Program (WAP)*, which is a voluntary cost-share program and is adapted to the individual characteristics and goals of each local farming operation. Farmers receive funding to implement structural improvements, environmentally friendly equipment purchases or additional management assistance (Isakson, 2002). Each WAP participant undertakes a process with local management bodies to create a personalized Whole Farm Plan (WFP), which specifically recognizes and manages source contamination within the operation. The Whole Farm Plan is organized by a Farming Team, which includes the landowner, local farming organization and government agency representatives (Isakson, 2002). The process of the WFP also includes selecting BMPs to manage potential sources of contamination which are also consistent with each farmer's operational objectives and limitations (Isakson, 2002).

The local watershed management approach in the Catskills has been greatly successful with an 85 percent rate of participation. This is largely due to the sense of ownership built by landowner participation and the funding provided to participants. As well, success of the WAP is largely due to the flexibility and adaptability of design for different categories of farming operations and the recognition of the financial benefit when implementing BMPs. For example, some farms did not meet the criteria, but are still considered potential producers of non-point source contaminants. Therefore, a Small Farm Program (SFP) was developed to provide funding incentives for small farming operations interested in implementing water quality and environmental management techniques (Isakson, 2002). The NYC management program is similar to the approach promoted by Ryan at Ontario's Grand River Conservation Authority (GRCA), who suggest that BMPs should be divided into two distinct categories for implementation: commercial operations and hobby or smaller farming operations (Agnew, 2005).

The second program introduced from the Watershed Management Plan includes the Conservation Reserve Enhancement Program. The purpose of this program was to pay farmers for voluntarily retiring sensitive parcels of land on their operations (Isakson, 2002). The third program introduced from the Watershed Management Plan includes the Whole Farm Easement Program (WFEP), which offers compensation for farms who decide to abstain from their development rights in areas considered to be environmentally sensitive.

This program has been quite successful in instituting a solid framework for watershed-based environmental management and it is perceived to be non-threatening by the agricultural community. For a program such as the NYC's Watershed Management Plan to be a successful in Ontario, there needs to be much greater financial commitment from the federal and provincial government than has been the case in Ontario (Agnew, 2005).

The NYC's Watershed Management Plan is only practiced in the New York State and this may not be representative of the situation in the US. A program of national spread like the Environmental Quality Incentive Program (EQIP) might be a quality assurance program that will aid Canada in developing a comprehensive environmentally sustainable program. EQIP is part of the Farm Security and Rural Investment Act, which was developed by the United States government in 2002. The purpose of the program is to provide a voluntary conservation program for farmers, so that they comply with national goals. It also provides financial support through program incentives, and also provides assistance to farmers to install environmentally responsible, management practices on their farm. Farmers must sign contractual agreements that hold them to program participation anywhere from 1 to 10 years, during which time incentive payments are provided on a cost-share basis of up to 75 percent. The U. S. Farm Bill directly supports 60 percent of the funding and the program is administered by the Natural Resources Conservation Service (NRCS). Recently the NRCS created a set of 24 action items to place emphasis on livestock operators' practices (Agnew, 2005).

There are also other programs such as the Milk and Dairy Beef Quality Assurance Program, a national voluntary assurance program for the American dairy industry, the On Farm Assessment and Environmental Review (OFAER) developed by the American's Clean Water Foundation (ACWF) to improve the overall environmental stewardship of pork producers and for protection of surface and groundwater.

CONCLUSION

Montpetit (2003) has recently observed that, unlike France and even to a lesser extent the United States, Canada does not coordinate its financial aid to farmers with agri-environmental regulation and this partly explains its relatively weaker agro-environmental effort. Since the publication of Montpetit's book, the Ontario has moved tentatively into the area of regulation by implementing the Nutrient Management Act, the Greenbelt Act and the Clean Water Act. All of these Acts add significant regulation to the voluntary programs that Ontario farmers have had access to over the years ranging from the Environmental Farm Plan to the Rural Water Quality and Clean Water Programs.

The development of the environmental programs in Europe appears more advanced than most Canadian programs. The European experiences indicate that voluntary programs work best when people already have significant incentives to change their environmental farm management. As well, voluntary programs have been more successful in reducing the impacts of pollutants that have a cost, such as pesticides, and does not work as well for items like nutrients and animal waste, where moving the material off-farm may cost the farmer. The EU is focusing on competitiveness, environment, fair income, simplified legislation and decentralization. As well, the EU is concentrating on increasing the availability of funds for specific themes and geographic locales. EU officials have found success by working with farmers and trying to understand their motivations in different context. For example, the Dutch adopt intensive agricultural practices, but have implemented many stringent policies aimed at specific effects to the environment, such as manure management.

Payments for environmental services first started in the EU and have since been adopted all over the world. Many of these programs have been adopted in the U.S. at the local, regional and state level. New York City's Watershed Management (WAP), for instance, is a voluntary cost-share program and adapts to the individual characteristics and goals of each local farming operation. Farmers receive funding to implement structural improvements, environmental friendly equipment purchase, or additional management assistance.

One reason why voluntary programs in the U.S and Europe have been effective is the funding they have received to allocate to farmers, technical support, equipment, training programs and public awareness. Policy makers and EU government officials in particular have been careful to consider the scale of operation and the type of farmer.

Though initially brought in as a defensive maneuver to avoid more direct government regulation, the EFP is one of the most effective programs in Ontario since it works with individual farmers to identify environmental risks on their farms and then develops solutions that will provide maximum protection for the environment. As well, the program permits farmers to contribute their own labour and equipment as part of the agreement, which will help farmers gain confidence in how the new program is administered. The main goal of the Ontario Canada EFP is to allow farmers to continue to be leaders in environmental stewardship.

A significant contradiction continues to exist between government departments implementing regulations and farmers wanting self regulation. Another contradiction exists between what most urbanites want and farmers will put up with such as over the Greenbelt Act, the Species at Risk Act, Normal Farm Practices, etc. At this time, Ontario has chosen not to participate in the group EFPs (unlike in Saskatchewan for example) but Ontario farmers still qualify for NSFP funding.

In Ontario the OSCIA, in concert with OFEC, has combined the Greencover Shelter, Farm Stewardship and Water Expansion funds together into 36 BMP options for which Ontario farmers can receive federal and provincial funding for environmental programming. In the aftermath of the CWA, an exemplary program has been developed by nine CAs within southwestern Ontario, who have come together to complement funding available through the EFP to provide incentives to adopt a variety of BMPs in the Clean Water Program.

There is growing support among the members of farm organizations for direct payments to farmers in return for providing environmental goods and services (EGS) like the retirement of fragile land, improved fuel storage and the protection for wetlands (Environics Research Group, 2006). There are a growing number of examples of Canadian funding for EGS such as Alternative Land Use Services (ALUS) but these pilot EGS programs still fall short of what the EU has been providing its farmers because the funding is not yet sufficient and still follows a market instead of a social and environmental rationality. Farm strategies for survival, especially for the smaller producers, may hinge on a combination of a revised Canadian Agricultural Income Stabilization and a new form of EGS which could benefit substantially from a close examination of the EU's Common Agricultural Policy and we will return to this prospect in Chapter 7.

In: Agricultural and Environmental Security...   ISBN: 978-1-61668-156-2
Editor: Glen Filson   ©2011 Nova Science Publishers, Inc.

Chapter 4

FACTORS AFFECTING FARMERS' ADOPTION OF BEST MANAGEMENT PRACTICES IN ONTARIO'S GRAND RIVER WATERSHED

Glen Filson, Delia Bucknell and Stewart Hilts

INTRODUCTION

To see how recent regulations and programs have operated in the Grand River watershed, we talked to Conservation Authority officials and consulted recent water quality studies to identify the sub-watersheds with the worst and best water quality. This chapter describes how residents within the sub-watersheds of the Grand, the Eramosa/Speed (E/S) and the Canagagigue Creek (CC) perceive agri-environmental programming and recent legislation imposed after the Walkerton tragedy. Earlier research had found fewer water quality problems in the E/S region than in CC (see GRCA, Feb. 14, 2002; Beak International *et al.*, 1999). We thought this might be due to different land use practices, degrees of agricultural intensity, the attitudes of rural people toward environmental best management practices (BMPs) and the impact those practices might have on rural people's perceived quality of life.

In keeping with the conceptual framework, we recognized that the adoption of BMPs is an environmentally responsible behaviour arising out of environmental concern. "BMPs are proven, practical and affordable approaches to conserving water, soil and other natural resources" (MVCA, 2005). A comparison of residents' views about the voluntary adoption of

BMPs versus the relatively new requirement that large operation farmers implement a Nutrient Management Plan (NMP)[2], enhances our understanding about whether farmers were adopting BMPs because of their environmental concerns or primarily because it was required of them.

To obtain a clearer picture of farmers' views about regulation respondents were also asked to comment on regulatory aspects of the Nutrient Management Act (NMA) (2002) and to say what they thought about voluntary environmental cost share programs like the Environmental Farm Plan (EFP) and the Grand River Conservation Authority's Rural Water Quality Program (RWQP). The survey was conducted with both farm and non-farm residents in these regions to obtain their perceptions about the effects of their local farming systems and whether they suspected large, medium or small farm operations of being the biggest causes of pollution. We tried to determine whether we could identify different levels of adoption of BMPs as a function of their farm characteristics, the reported water quality of particular sub-watersheds and/or their demographic characteristics.

FACTORS MOTIVATING FARMERS TO INTRODUCE BMPS

As early as 1991, Duff *et al.* noted that education and financial incentives are seen by farmers to be more effective than other government policies in promoting conservation. They observed that farmers tend to prefer a voluntary approach to environmental management, despite the fact that they perceive regulatory approaches to be more effective in some circumstances. Kaiser and Shimoda (1999) also observed that people feel morally responsible more as a result of guilt feelings than for the reason that they must behave conventionally.

Cestti *et al.*, 2003 argued that the effectiveness of BMPs designed to improve environmental conditions is a function of local conditions including 1) topography; 2) climate; 3) cropping system; 4) maintenance; 5) proper site selection and 6) proper installation. Thus the effectiveness of BMPs has a lot to do with site specificity. This also affects farmers' motivation to adopt because if a program is not tailored specifically enough, farmers may not be

---

[2]Relaxing the earlier requirement that all farmers introduce an NMP occurred in part because of the widespread perception that environmental pollution is mainly caused by so-called 'unsustainable factory farms' but the Walkerton tragedy occurred as the result of an improperly chlorinated well on a small to medium sized beef farm which had done the due diligence of implementing an Environmental Farm Plan.

interested, and this too becomes a barrier. This is also expected because there are different types of mindscapes (Maruyama, 1985). These different types of mindscapes affect the landscapes and the ecological goods and services produced by farmers as do their farm sizes, the types of commodities they produce and their overall economic well-being.

Intention is also very important in addressing the issue of who adopts environmental best management practices. According to the Hines Model (cited in Fransson and Garling, 1999), intention is a factor most closely related to actual behaviour. Intention is related to knowledge, skill and personality. Thus pessimistic farmers will only adopt best management techniques if they have the intention but not necessarily because of regulation. Intention can be developed by interacting with farmers who had already adopted the environmental management practice. However, some farmers probably implement BMPs because they are being told to do so (Filson, 1996) or receive peer pressure from family and friends to do so (Ryan, 1995).

Farm characteristics influence the adoption of voluntary conservation practices and programs in several ways. According to Gale *et al.* (1993), the following factors can have an effect on whether farmers will adopt conservation measures. First, the larger the farm, the more likely and more willing farmers are to adopt new practices, as larger farms require more information to operate and would therefore be more exposed to environmental literature and issues (Tucker, 2002 and Serman, 1999). Second, the higher the sales and income on a farm, the more likely programs will be adopted (Luar, 1993) as larger farms generally have higher incomes, and smaller farmers may be unable to commit what could be a significant proportion of their incomes to conservation measures. Third, it is hypothesized that there is less priority toward rented land and the reception of conservation program (Gale *et al.,* 1993). Fourth, the type of farm enterprise will affect the adoption of conservation practices (Morris and Crawford, 2000).   Gale *et al.* (1993), for instance, found that farmers with lower economic status had lower participation rates in the American Rural Clean Water Program (see chapter 6 below).

Also, farmers' involvement in environmental organizations means they will be more likely to adopt an environmental program (Smithers and Furman, 2003; Luzar, 1993 and Morris and Potter, 1995) as they will likely be more exposed to the effects of environmental conservation programs. Involvement in environmental organizations also affects farmers' cognition as the result of interaction. Belonging to such organizations may affect farmers' locus of control which in turn may affect their adoption of best management practices.

Farmers with an internal locus of control may be better adopters and environmental managers.

Religion can also play a role in farmers' environmental consciousness and therefore affect their likelihood of adopting conservation practices. While researching the Clean Up Rural Beaches program, Ryan (1999) found that farmers belonging to the Christian Farmers Federation of Ontario (CFFO) had lower environmental attitude scores than Ontario Federation of Agriculture (OFA) members, who in turn had lower scores than lifestyle farmers. She argued that Christian theology may play a role in farmers' environmental attitudes. She noted that "The statements about the finite resources of the earth and the relationship of humans and nature appeared to cause the most concern" (Ryan, 1999: 194). After observing that the CFFO's *Earthkeeping Ontario* newsletter and policy statements reflect respect for the environment she argued that leaders may be more progressive than their members.

Stewardship or 'earthkeeping' implies a use of creation to benefit all people and creatures. The concept is that if we care for creation it will provide us with all we need to survive. This differs considerably with standard environmental ideology that tends to emphasize preservation rather than use (Ryan, 1999 194).

Stern and Oskamp (1987) also assume that activated norms are sometimes accepted in more specific form as norms for one's behaviour. A violation of these personal norms involves feelings of guilt whereas compliance invokes feelings of pride. Religion is a very important determinant of the values and norms that prevail in any society.

Former CFFO strategic policy advisor van Donkersgoed has called for a suspension of the NMA and other regulations affecting Ontario farming (2005). While the CFFO has agreed that large farms ought to "document their stewardship" (van Donkersgoed, 2004) the CFFO feels that the rest of society should pay farmers for producing "our clean water, fresh air, biodiversity, abundant wildlife and attractive landscapes" in the form of "annual whole farm payments" (van Donkersgoed, 2005) which is what European Union farmers get in return for producing Environmental Goods and Services (EGSs) (see chapter 2 and 3).

Serman's 1999 analysis of farmers' perceptions of Ontario's 1980s and early 1990s Soil and Water Environmental Enhancement Program (SWEEP) found a correlation between program uptake and farm size, gross farm income and farmers' educational levels. Several other Ontario studies have been done to try to understand farmers' motivation to adopt BMPs as a function of farm and demographic characteristics (e.g. Stonehouse, 1996; and McCallum,

2003). Economic constraints, mistrust of government, insufficient promotion, a shortage of skilled agricultural extension workers and the perception that the environment is fine as it is were among the most commonly cited reasons why some farmers do not participate in conservation programs.

In 2001, southern Ontario's Maitland Valley Conservation Authority (MVCA) commissioned a study of landowner motivations particularly among its concentrated livestock producers and found relatively positive attitudes among farmers towards the introduction of conservation practices with the vast majority of interviewees agreeing with the need to fence cattle off from streams and prepare Nutrient Management Plans (Kayak, 2001). The survey found 78% of farmers willing to adopt environmental programming if there were financial incentives like cost-sharing and tax reductions with technical advice and resources to facilitate program implementation (Kayak, 2001).

Blackie and Tuininga (2003) investigated complaints in Ontario's ten most threatened watersheds resulting from manure spills recorded by the Ministry of the Environment's Spills Action Centre. The top three watersheds impacted by livestock spills and discharges were the Ausable, Grand and Maitland Rivers. Ausable Bayfield watershed had the most manure spills in Ontario with 29% of its existing manure storages undersized. Such manure management problems as winter manure spreading and failing to check tile drainage for manure discharge were common. The larger, commercial livestock producers generally had the most appropriate manure handling facilities whereas substantial underinvestment in such facilities was the norm for most of the smaller operations. Obviously there is a need to improve on the use of BMPs to not only reduce excess nutrients but to minimize pesticide use, prevent soil erosion, support biodiverse environments and limit the production of greenhouse gases such as methane produced by livestock.

One of the GRCA's most important innovations has been the development of its Rural Water Quality Program (RWQP) (reviewed in chapter 3). This cost-share program funded mainly from local municipalities with help from higher levels of government, enables farmers to qualify for funds to improve their manure storage, fence cattle from streams and introduce other BMPs like riparian buffer strips. Farmers can qualify for grants for manure, fertilizer and fuel storages, help with the development of nutrient management plans and the retirement of fragile lands. Grants can be between 50 to 100% of the cost, depending on the kind of project funded (Schultz et al. 2004). Since inception in 1998 over $16 million had been spent by the RWQP and more than $8 million by farmers on such projects (Ryan, 2006) and the amounts continue to climb.

The CC has been selected as one of five surface water priority areas in the Waterloo Region where farmers can qualify for RWQ grants (Gray *et al.*, 2004). *The Canagagigue Creek Watershed Report* (2002) discussed several issues connected with this study including erosion of watercourses through animal access, lack of vegetation along the sides of these watercourses (vegetative buffer strips), and surface run-off. This has contributed significantly to the Creek's poor water quality. Programs such as the Rural Water Quality Program, developed by the GRCA's Ryan with other rural stakeholder groups, have been successful to a degree in mitigating some potential for risk, though problems still exist.

The GRCA's Ryan (1999) has argued that there are both commercial BMPs, like manure storages which will eventually make farmers money, and purely environmental BMPs like fencing which come at a cost but yield no financial benefits. Farmers have more incentive to introduce the former but the RWQP helps farmers introduce both types as long as they have first completed an EFP. While Diffusion of Innovation theory helps explain why the commercial BMPs are introduced by innovators and early adopters, who usually operate bigger farms, have higher incomes and educational levels and are less bound by tradition, as predicted by Rogers (1995), Ryan calls for a Theory of Altruism to explain why some farmers introduce net cost BMPs in a manner often inconsistent with the characteristics that Rogers argued usually predict innovators and early adopters.

Trust between farmers and the environmental agent, whether it is with the government, an environmental organization or non-farm rural resident, is not easily earned. It is important for farmers to see that agents are willing to make significant commitments, whether it be funding or program longevity for example, as it is unrealistic to expect farmers to do the same if this commitment is not reciprocated. If a climate of reciprocity is not obtained in the form of commitments, farmers will be unlikely to trust the program that is being proposed or implemented (Marshall, 2004), and will therefore be unlikely to adopt a best management or other conservation practice.

Rogers (1995) found that information obtained from peers carries more weight than information from the media, scientists, and likely the government. Yet it is important that farmers feel they can trust the government, which will increase compliance with any programs implemented (Marshall, 2003). Ensuring that farmers are consulted, and that enough funding is available to ensure compliance will provide the building blocks of trust between farmers and the government or other agent to foster healthy stewardship practices.

The attitudes of farmers and non-farmers in a relatively good water quality sub-watershed of the Grand River (Eramosa/Speed) were compared with the attitudes of farmers and non-farmers in a sub-watershed which had very poor water quality (Canagagigue Creek) to see if there were systematic differences.

THE STUDY AREAS

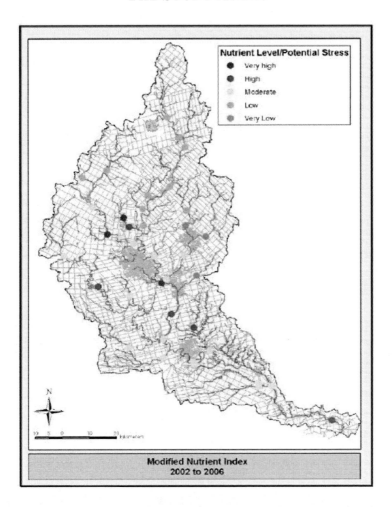

Figure 4.1. Nutrient Index integrating nitrogen and phosphorus concentrations at selected sampling sites in the Grand River watershed (2002-2006), Sandra Cooke, GRCA)

Located on the outer edge of the Greater Toronto Area (GTA), the Grand River watershed has its own urban pressures from the cities of Guelph, Kitchener-Waterloo, Cambridge and Brantford. About 800,000 people live in the watershed in part because of its prime agricultural land, favourable climate and access to markets (Schultz *et al.*, 2004). In the figure above, the E/S sub-watershed, with good water quality, passes south into Guelph whereas the CC, with poor water quality, passes from Elmira south into Kitchener/Waterloo. As the Speed River passes through Guelph, inadequate storm water treatment causes a decline in water quality but CC water quality is generally much worse than that of the E/S. This motivated us to compare the views of rural people in each area to environmental issues, BMPs and the impact of intensive agriculture on people's quality of life.

By contrast with the E/S, the area surrounding the CC is primarily rural, with a significant number of farms. The average age of farmers interviewed in the Canagagigue area was 49.3 years, with all of the respondents being male. This is similar to the average age of the farming population as a whole in the area though the average age of farmers throughout Ontario is even higher. The average farm size in the study was 117 acres and ranged from 31 to 270 acres with a median of 100 acres, which is smaller than the average of approximately 167 acres in C.C. overall. With respect to level of education, 72% of the farmers had completed grade eight, 14% had completed some of high school, and 3% had attained a diploma in agriculture. The average educational attainment in the area as a whole was higher with 30.1 % of the farmers having a grade 8 education and 26.1% having completed high school.

Farm operations that had beef cattle as a commodity were the most prevalent, with combinations with either pork or dairy being common. Poultry was the least common of all livestock operations, with only three of the farms surveyed having any poultry, whether stand alone, or in combination with other animals.

Within the Creek several factors damage water quality. Erosion is evident in 62% of CC's stream banks exacerbated by increased pasturing, tillage and cultivation along watercourses. Eutrophication of reservoirs is of concern as well, because as phosphate levels continue to increase, levels of dissolved oxygen decrease, in turn hurting oxygen-reliant organisms (GRCA, 2002).

Both point and non-point source pollution in the Creek have caused environmental problems. In 1989 it was discovered that a shallow aquifer near Elmira had been contaminated with nitrosodimethylamine and cholobenzene, as a result of leaching wastes from Uniroyal, a major chemical manufacturing company (Guelph Water Management Group, 2005). The result has been indefinite closure of many source wells and aquifers (Schultz *et al.*, 2004). Instead of cleaning up their DDT and dioxin laden-site, the company had been fighting efforts to force a clean-up to the detriment of the Mennonite farmers surrounding the plant. "Mennonites tend not to engage politically in land use and other decision-making and consequently suffer for their beliefs" (Dempster *et al.*, 2004: 157) because they often chose not to demand that Uniroyal clean up their pollution.

The GRCA's Water Resources Manager, Lorrie Minshall also indicated that more than half of the water quality problems in the CC are due to non-point source pollution, largely from agriculture. Run-off from fields, barnyards and improper manure handling and storage are among some of the issues that must be addressed. Other issues include animal access to stream banks and loss of topsoil, representing significant non-point sources of pollution (Minshall, Oct. 3, 2002). In 1999, the CC was identified as a high priority for the reduction of agricultural pollutants. The following livestock farms were considered high priorities for the RWQP (table 4.1).

Table 4.1: Non-Point Course Inventory of Canagagigue Creek Problems (Source: GRCA, Feb. 14, 2002)

Total Livestock Farms: 200 dairy, 78 beef, 25 swine, 19 poultry, 8 horses	315
High priority manure storage needed	109
Medium priority manure storage needed	151
Low priority manure storage needed	55
Projected milkhouse impacts	66 (1/3 of dairy farms)
Livestock access sites to the Creek	140
Tillage in floodplain	13
Sites with visible soil erosion	293
Sites requiring buffers	272

Comparisons of Eramosa (within E/S) and Peel Townships provide useful insight to the differences between E/S and CC. Peel has been amalgamated with Maryborough and Drayton Village to form Mapleton Township. Generally the landscape in the Eramosa region is more rolling, with drumlins and wetlands in lower valleys, and a higher proportion of remaining forest cover. In the former Peel Township, in the upper CC watershed, the landscape is much more flat, with a very low proportion of remaining forest cover.

CC farms are considerably more intensive, in terms of both crops and livestock, than in the E/S region. In a study of the biodiversity within the region, Okey (1998) found a relatively high density of livestock in the CC sub-watershed compared with other parts of the Grand River watershed. According to Statistics Canada (2001), in Mapleton there are 491 farms that carry cattle and calves; 336 farms in Woolwich and 221 farms in Centre Wellington that carry cattle and claves. Pigs are also raised a lot in Mapleton with 159 farms; 138 farms in Woolwich and 69 farms in Centre Wellington. There are also 40 farms in Centre Wellington that raise sheep and lamb; 36 farms in Mapleton and 26 farms in Woolwich. Other livestock include variables such as horses and ponies-where Woolwich has 223 farms that raise horses and ponies; Mapleton has 187 farms and Centre Wellington has 92 farms. Lastly, the poultry inventory includes the number of farms that raise chickens. Mapleton has the greatest number of farms that raise hens and chickens with 227 farms; Woolwich has 207 farms and Centre Wellington has 74 farms that have chickens. Overall, Mapleton seems to concentrate largely on livestock and poultry compared to other areas (Statistics Canada, 2001).

Peel Township (now Mapleton) has 83% of its farmland in crops while Eramosa Township has only 62.5%. Conversely, Peel has only 7.9% of farmland in 'Other' (mostly forest cover), the lowest in the county, while Eramosa has 22.3%, the highest. Peel Township also has intensive livestock farming, with cattle averaging one for every 2.5 acres, and pigs one for every 0.87 acres, while in Eramosa these averages are one per 4.0 acres, and one per 2 acres. This results in an even larger difference if manure production from this livestock is considered. Whether measured in terms of biological oxygen demand, nitrogen loading, or phosphorus loading, Peel Township exceeds Eramosa Township by a ratio of 5:1.

These fundamental differences in the agricultural landscape underlie the differences in farming practices and in attitudes found in this study. The CC is simply far more intensive agriculturally, despite their relatively small average farm size, than E/S especially with respect to livestock.

By contrast, Beak *et al.*, 1999 reported that the Eramosa River and its tributaries have high water quality with the Speed River having only slightly worse quality. About 45% of the land is agricultural. The improvement of the environmental quality of the area was attributed to the conservation ethic of landowners and their support for programs conducted in the region and the unique physiography, hydrogeology and vegetation (Bucknell, 2002).

Of the 11,065 people within the Guelph/Eramosa Township 90% are English only, 1% French only and over 8% many other language groups. Thirteen percent were foreign born. About 50% are Protestants, 30% Catholic and remaining 20% are either not religious or are 'other Christians,' Jews, Sikhs or Buddhists (Statistics Canada, 2001).

CC lies mainly within Woolwich in Waterloo Region, Mapleton and Centre Wellington Townships (formerly Peel and Pilkington). Of the 27,060, including those who live in these townships and towns, about 75% were English only, 28% had a language other than English or French. German Canadians are next followed by the Dutch. Seventy percent of these regions are Protestant, 16% Catholic and 14% belong to some other religion or none at all. A sizeable percentage of the Protestants whose first language is not English are Mennonite (Statistics Canada, 2001).

The C.C. region is unique in that much of the land is dedicated to agriculture and that a significant number of Mennonites reside in the area. In the CC area there are 550 Mennonites in Centre Wellington; 2,990 Mennonites in Mapleton and 4,270 Mennonites in Woolwich (Statistics Canada, Sept. 2001).

There are at least 18 separate groups of Mennonites within Ontario. Approximately 66% are within Waterloo County where Canagagigue Creek exists (Wandel, 1995). From the early 1800s Old Order Mennonites (OOM) have been settling in what is now Mapleton, Wellesley or Center Wellington Township. Between 25 and 35% of the population in Waterloo County are Mennonites (Wandel, 1995), while the Mennonite population within Mapleton and Woolwich Townships specifically, is approximately 24% (Filson, 2004). The area includes conservative Old Order Mennonites (David Martin, Elam Martin and Old Colony), moderate Mennonites (for example, Evangelical and Beachy Amish) as well as progressive Mennonites (e.g. Mennonite Conference) (Fretz, 1989).

Typically they are committed to a 'plain' life of hard work, thrift, pacificism and a reluctance to be involved with government (and hence government programs like the RWQP). The diversity of groups in the Mennonite faith vary from the Old Order Mennonites (OOMs), where no electricity, no automobiles and strict separation from the outside world is enforced, and so-called Black Bumper Mennonites, who may drive a black car to church, though still remain relatively traditional, to the other end of the spectrum where Mennonites are indistinguishable from other farmers or citizens. For this latter reason, it has become essential for GRCA conservation stewardship extensionists to carefully explain the functions of the program to the Mennonite religious leaders because in the first years of the RWQP there was very little uptake due to the GRCA's association with municipal government. About 75% of the RWQP have been done by the Dave Martins group whereas some OOMs, who will do the EFPs (farmer based), will not take the RWQP money because of the government funding which is attached (interview with Tracey Ryan, Oct. 3, 2002).

The core area of Old Order Mennonites and other Mennonite groups live in the western part of the Grand (see the map above) where they constitute between 25 and 35% of the population (Wandel, 1995). Thus a strong OOM culture (Arthur, *et al.*, 1998) continues in the CC area though there are many non-Mennonites as well. Eramosa, on the other hand, has a broader religious and cultural base (Statistics Canada, 2002). The presence of the Old Order Mennonite group in the CC area was a factor in the research and was considered an important socio-cultural factor helping explain the findings (see the second phase of this research below).

RESEARCH METHODS

The data for this study were collected initially by conducting two focus groups, one with farmers, the other with non-farmers in the E/S region. The classification of land in the Municipal Assessment Roll was used to indicate whether or not a person was a farmer. Questionnaires were then completed via mail survey in 2002 by 385 randomly sampled farmers and non-farmers within the E/S and CC sub-watersheds of the Grand River. The questionnaire was based on the issues raised in the focus groups, the extent of use of BMPs, environmental attitudes and the impact of farmers' practices on perceived quality of life. Frequency distributions of the responses to questions were calculated along with inferential non-parametric statistics.

Relatively lower uptake of environmental BMPs, combined with our knowledge that CC sub-watershed is the Grand's more severely degraded sub-watershed, led us to conduct phase two: 29 further face-to-face, telephone interviews and an additional survey in that region. Most of these respondents wore black clothing but the questions were not about religious issues but instead, their degree of participation in EFPs, BMPs and nutrient management.

The sample in the E/S watershed was collected from the Municipal Assessment Roll at the Eramosa Township office using a random selection process. Individuals were included in the sample if they fit the criterion of living within a rural area in the region. Six hundred surveys were sent out in the E/S region, 300 to farmers and 300 to non-farmers, in 2002 but some were returned and others felt they did not qualify, leaving the total sample at 585. Thirty-two percent responded including 66 farmers and 120 non-farmers. This under-represents female farmers and may over-represent those most adept at completing questionnaires who have with relatively higher educational and income levels.

The sample consisted of 53 women and 130 men including 9 farming females, 44 non-farming females, 55 farming males and 75 non-farming males. Most respondents in this region had completed high school and there was no significant difference in the education of the farming and non-farming groups. The mean age of all the respondents was 55.9 years with a range from 26 years to 91, which is significantly older than in CC but more typical of Ontario's farmers as a whole. Most respondents had gross incomes of less than $39,999.

A total of 651 surveys were sent out in the CC sub-watershed (minus the town of Elmira); 18 of these were returned or were received by people who did not feel they were qualified to respond leaving a total of 633 with a response rate of 31.5 percent. The sample consisted of 86 (43.2%) farmers and 133 (66.8%) non-farmers. The gender division was similar to that of the E/S study with 60 females, 133 males and 6 who did not indicate their gender. The education level of respondents, both farmers and non-farmers, in CC was significantly lower than in the E/S sub-watershed. The mean age in CC was 50.6 years, and the range was larger with ages of from 12 years to 94 years. Surprisingly a significant difference in income levels was seen between farmers and non-farmers in this region with non-farmers tending to be in the lower income brackets. Though most of the farmers were in the lower income bracket, farmers had higher incomes on average than the non-farmers.

Based on the size of the sample for the total rural populations of the two areas, the response rate from the two regions for the questionnaires had sampling errors of about 7% in CC and 5% in E/S but the sampling error is higher for specific groups such as female farmers. Given the nature of the data, variable partitioning using discriminant analyses was ruled out in preference for nonparametric statistics which require the fewest assumptions about the findings. Non-parametric statistical analysis was used because they do not require perfect probability sampling techniques, normally distributed data and the use of continuous variables.

In the first phase, farmers were asked if they would like to receive a summary of the research. When they received the summary results they were asked if they would be willing to participate in face-to-face interviews regarding government regulatory policies and environmental practices on their farms in exchange for soil tests of their land. Five farmers were interviewed in person and 18 were interviewed by phone. Help from the director of the Mennonite Cultural Centre in St. Jacobs generated an additional six surveys with farmers and a detailed soil tests were completed for interested farmers.

The mean age of the smaller group selected for intensive interviewing was 49.3 and all were male. Farms among this latter group were 30% smaller (117 acres) compared to the average of 167 for C. C. as a whole (Statistics Canada: Agricultural Census 2001). Only 3% had an agricultural diploma, 11% had completed high school, 14% had some high school education, while 72% had completed grade eight only. Most farms were small mixed livestock (beef, dairy, pork and to a lesser extent poultry) and cropping operations with grains and oilseeds followed by winter wheat, fruit and vegetables.

The main objective of the second phase of the study, conducted in the winter of 2004, was to study C.C. farmers and non-farmers' attitudes and beliefs regarding environmentalism in the C.C. sub-watershed of both farmers and non farmers. It also sought to assess farmer attitudes toward environmental programs and legislation including the NMA, EFP and BMPs. We studied the effects of the new regulations on farm operations in addition to compiling a social profile of farmers and their farms in C.C. This was particularly important due to our small sample and the need to determine whether the sample was reasonably representative of farmers' demographic and agricultural operations in the area. The social profile was used to aid our understanding about the characteristics of the study population and explore cultural differences about adoption, or lack thereof.

For the follow-up interviews conducted in 2004, a sub-sample was taken from the 2002 sample. The original farmers were contacted from a reply card from the study completed by Bucknell in which they were asked whether they would be willing to be contacted to gain further information about the results of the study. Five of the farmers agreed to being interviewed at their farms in November of 2003. Other farmers were contacted by telephone from a list which was provided from the original study asking about their farming and environmental practices. A third attempt to increase the sample size was made when the Director of the Mennonite Cultural Centre in St. Jacobs handed out surveys, with six farmers returning the surveys. The total number of farmers in the sample was 29, while the number of non-farmers interviewed was eight. The latter were representatives of environmental organizations, conservation authorities and non-farm rural residents.

Along with her return of the uncompleted questionnaire, one OFA member asked "Do Mennonite people of any sect take part in such surveys? No!" She then went on to provide a nice overview of the issues discussed in the questionnaire and concluded by saying:

"If there is a place for 'ecological goods and services' wherein the public can partake, without adding to the tax bill, without benefiting those who take without giving in society and whose mandate is clearly stated, understood by all, then we would probably support it. We do believe in having our land remain productive, with little or no erosion, in having clean water. We use a cistern and roof(s) rainwater for almost everything but drinking and will continue to do our best to follow anti-pollution rules and regulations."

The findings from the in-depth interviews, conducted to learn more about the specific issues for this group with the high percentage of Old Order Mennonites are summarized below (see Wells, 2004).

MORE FINDINGS

The study assessed whether respondents in both sub-watersheds felt that their perceived quality of life was influenced by the environment as well as agricultural policies such as the Nutrient Management Act (2002). Non-farmers in both Eramosa/Speed and Canagagigue Creek regions tended to find the scenic quality of the environment to be more important than did their farming counterparts (E/S: $\rho = 0.197$, $p \leq 0.008$; CC: $\rho = 0.159$, $p \leq 0.025$). Non-farmers within CC felt that environmental quality was more important to their quality of life than did farmers ($\rho = 0.144$, $p \leq 0.044$). Also, non-farmers in both regions felt negatively about air pollution created by other peoples' activities. In the E/S region non-farmers were also more likely to indicate that they were somewhat negatively affected by both air pollution and by water pollution from other landowners as well as their own activities.

The Freidman test was used to ascertain whether there were any differences between farmers and non-farmers in the amount of impact on their perceived quality of life from any one type/source of pollution. The variables tested were: water pollution from others, water pollution from own activities, air pollution from others, air pollution from own activities, soil erosion/degradation, and soil compaction. In both regions farmers were more concerned about the soil and water pollution from others and non-farmers were also more concerned about the air and water pollution from others.

Respondents were asked to indicate their level of agreement on a five point scale from strongly agree to strongly disagree with thirty-one statements dealing with agricultural and environmental issues. All of these variables were tested using Spearman's rho correlation. More non-farmers agreed with the environmentally conscious statements than farmers and those who were younger, female and had a higher education also tended to be more environmentally conscious. This result is consistent with the findings of the review of literature conducted by Fransson and Garling (1999), although the impact of gender on environmental concern was ambiguous. Arcury and Christianson (1990) suggest that men are more concerned than women while Stern *et al.* (1995) asserts otherwise as does Filson (1996). Gender differences may indeed be site specific.

Respondents' attitude towards the size of farm operation was gauged by asking respondents to indicate what they saw as more of a threat to the environment - many smaller farms, fewer larger farms, both or neither. In the E/S region, non-farmers and farmers agreed that both small and large farms are a potential threat to the environment while in CC both farmers and non-farmers agreed that fewer larger farms represent more of a threat (see Table 4.2 below).

Table 4.2: Size of Operation Perceived to be More of a Threat to the Environment

	Eramosa/Speed				Canagagigue Creek			
	farmer		non-farmer		farmer		non-farmer	
	Which type of operation is more of an env. threat?		Which type of operation is more of an env. threat?		Which type of operation is more of an env. threat?		Which type of operation is more of an env. threat?	
	count	%	count	%	count	%	count	%
many small farms	3	4.7	2	1.7	5	6.2	8	7.1
fewer larger farms	25	39.1	44	37.3	39	49.1	49	43.8
both problematic	26	40.6	58	49.2	31	38.3	45	40.2
Neither	10	15.6	14	11.9	6	7.4	10	8.9

Thus most respondents were relatively evenly split between those who feel that fewer, larger farms are more of an environmental threat than both small and large farms but few people thought that many small farms are as environmentally problematic as fewer larger farms, or a combination, are.

Similar results were found in both study regions on the topic of government regulation of agriculture. The E/S data suggested that non-farmers in that region typically favour government regulation of agriculture to protect the environment yet farmers would prefer not to have more government regulation of agriculture. The same tendency was found in CC with one difference; non-farmers tended to agree with farmers that 'government regulation is (not) the only way to ensure it works' (to protect the environment).

Farmers from both regions felt that government intervention in agriculture would affect their quality of life negatively; however, a majority of farmers still indicated that regulation regarding water quality could have a positive impact. The majority of farmers also felt that increasing the educational qualifications of farmers would protect the environment and that the added cost in both time and money was worth it, although they tended to disagree or were unsure as to whether government regulation was the only way to achieve environmental protection. Education is important because farmers become more environmentally responsible when they know the consequences of their actions. In contrast a higher percentage of non-farmers than farmers wanted to see increased government intervention into farming practices to protect the environment.

Both farmers and non-farmers felt voluntary cost share programs implemented by farmers such as the EFP were needed to increase environmental stewardship. The Kruskal-Wallis test of independence revealed that those who had participated in environmental programs were more likely to support 'soil testing in order to manage nutrient level' than were those who had not participated, so such interaction shapes one's self and affects one's cognition. Sixty-five percent of farmers said they had participated in an environmental program, and within the sample of farmers 45.5% had completed the Environmental Farm Plan (EFP) which, at the time was much higher than the county total (27.1%), or cash crop producers for Ontario (36%) who have a higher rate of EFP completion than most livestock producers and especially small, mixed farming operations (FitzGibbon et al., 2004).

The analysis of the environmental farming practices of the CC generally revealed a low uptake of BMPs, but a reasonably high degree of participation in environmental programs. Five more farming practices were adopted by those who had participated in environmental programs than by those who had not including the creation of tree windbreaks and block plantings, the employment of conservation tillage, the improvement of manure storage/handling facilities, development of erosion control structures, and the establishment of ditch and stream bank protection. Correlations were found with all of these practices that indicated that those who had participated in environmental programs were more likely to have employed these practices. However, the participation rate in environmental programs in the CC region was significantly lower than in the E/S region ($P \le 0.016$).

FOLLOW-UP INTERVIEWS AND SURVEYS IN THE CANAGAGIGUE CREEK (CC) AREA[3]

Almost half of the farmers (14) said there were no environmental problems in the CC area worth worrying about. Another 34% (10) said there were problems they should worry about and the remaining five did not know. When asked about what other farmers in the area were doing, 23 (79%) felt that farmers in the C. C. area were doing their part to maintain environmental sustainability, 5 (17%) believed that some were not and only 1 (3.4%) thought that farmers were definitely not managing their farms in an environmentally sustainable way. Of the 20 farmers with cattle adjacent to CC, 60% (12) fenced their cattle off from the stream while 40% (8) did not. Despite pressure to fence his cattle off from the stream, one farmer said that he would "fight it every step of the way" (Wells, 2004).

While some of these farmers were adamant that their neighbours were doing what they could reasonably be expected to do to protect the soil and water, others felt that they were doing "time honoured practices that their ancestors had be doing with no apparent ill effects" and a few were very opposed to growing pressure to change their practices. Most farmers were unable to identify any significant adverse effects that could be traced to their farming operations and so they were unwilling to participate in 'unnecessary' environmental programs. This is despite the fact that phosphorous and nitrogen production from manure are especially high in the C.C. areas at anywhere from more than 2 Kg/ha to 20kg/ha (Statistics Canada, 1996).

When asked whether there were any environmental problems in the area, or if they were worried about any specific problems, half of this study's farmers felt that there were no environmental problems to be concerned about. Another 33% felt there were some environmental problems, most commonly regarding water issues. One farmer mentioned drought, and four farmers

---

[3] For an in-depth review of the follow up interviews, see Wells (2004).

mentioned manure run-off. Seven people (24%) pointed to the former Uniroyal Chemical's serious pollution of the C.C. and others raised such issues as urban waste and automobile pollution. These environmental problems were seen as much more problematic than farmers' nutrient management. Regarding the barriers to their participation in environmental programs 38% point to the lack of money, 24% saw no barriers, 17% felt there was no need to change, 7% said there was too much regulation and/or problems with government run programs.

Though 41% of the farmers were unaware of any governmental conservation programs, 39% of the farmers interviewed had completed at least some stage of the EFP. Fifty-nine percent of the farmers had implemented at least one BMP while the others said they had not. Forty-seven percent had implemented windbreaks, 35% had manure storages, 5 had created buffer strips, 24% had improved their soil management and also improved their pesticide storage, 18% had implemented cleaner water management, 12% had fenced their cattle from streams, conducted intercropping and conventional tillage and 6% had done wildlife management and put in fence rows.

Regarding environmentalism, having motivation to adopt means having an incentive to do so, whether it be financial (tangible) or for stewardship reasons (non-tangible). Providing enough of an incentive for a farmer to make a behavioural change can be a difficult task as we are asking them to alter what could potentially affect their livelihood. The greatest motivational factor in farmers' decisions to adopt BMPs was that of seeing the need to make changes. Despite the water's poor quality, many farmers interviewed in this study sounded indifferent to the potential negative effects of farming on the environment.

Farmers were reluctant to see the benefits of implementing BMPs because they felt they were already being good soil and water stewards. It is important to note that two of the farmers admitted that they were "laughed at" by other farmers when they decided to implement environmental practices, so peer pressure can also have the opposite effect. Money, time, making BMPs mandatory and having more energy were all important in farmers' decisions to potentially adopt BMPs. The farming workday is long, and many farmers have off-farm employment, or other on-farm responsibilities in addition to their regular chores, which can provide little motivation. Many felt that by participating in environmental programs, more time would be diverted from income earning pursuits, and may result in a net economic loss.

Money is again a central feature in farmers' opinions regarding the adoption of BMPs. Farmers felt that without significant sources of funding such as grants from outside sources, implementing BMPs is too costly. Structures such as manure storage systems can be expensive, and without incentives available, adopting BMPs may adversely affect the farms' profitability and therefore their livelihoods. Some Mennonite farmers' unwillingness to accept grants for religious reasons is a fundamental concern, as they would be responsible for the entirety of the cost of the project, having a greater impact on the farm's profitability. Another barrier was that some farmers simply saw no need to change their farming practices (Wells, 2004).

Over half of the farmers relied on farming as their sole support system. Another 31% said that 75% to 99% of their income comes from farming and another 14% indicate farming accounts for 74% or less of their income. On average, the number of farmers in the study who work off-farm, or have other employment, was similar to those in the C.C. as a whole. Both men and women have increased their rate of working off the farm since 1990, and roughly equal proportions of women farm operators (45.6%) and men (44.2%) worked at non-farm jobs in 2000 (Statistics Canada, 2001). This factor within the study population demonstrates the need for a financial support for especially the least well off farmers to adopt BMPs.

The amount of work associated with the implementation of BMPs was also perceived by some to be a significant barrier. The paperwork involved, permits required and the time spent building or implementing the project were all perceived barriers. Several farmers were concerned that by adopting BMPs, their yields would be adversely affected. This could occur, farmers believed, if resources were diverted toward environmental conservation and away from the operation of the farm. Or, by creating buffer strips, they would also forfeit land for cropping. This raised farmers' concerns that these conditions might adversely affect their quality of life. Farmers felt that seeing the need for change is both a barrier and a motivation for adopting BMPs.

Support for environmental programs and for further government involvement was evenly split amongst the farmers. Half of the farmers' felt that the government was sufficiently involved with farmers' environmental practices as is, and another 37% felt that the government should be helping more, mostly through financial means like availability of grants and further informational support.

The EFP had been adopted by 39% of the farmers interviewed, which was somewhat higher than the average for the province. Farmers provided varying reasons for participating in the program such as to become eligible for grant money, observing peer adoption and concern about whether their farm operation was adversely affecting the environment. Another 60% had not started the EFP either because they felt it was unnecessary or that by doing it, they might lose control of their property, or in some measure be behaving inconsistently with their faith.

The NMA was a contentious issue for only a minority of farmers in the study. Few of the farmers interviewed had implemented a NMP, and most respondents had smaller operations with fewer than the 300 nutrient units at the time they were required to complete a NMP. Two-thirds of the farmers interviewed said the NMP was a good idea, however, the remaining third felt that they would only implement it if was made mandatory by government regulation for them as well. Still others felt that it would not affect them at all, as their operations were too small to ever fall under the NMA requirements.

DISCUSSION

The E/S region has less intensive farming including lower livestock densities so the residents experience less pressure to improve their water quality. Within CC, since Mapleton township carries a large portion of farms, this area also tends to have more livestock and Woolwich has the most crop production. Most of the rural communities are in the Mapleton and Woolwich areas where the numbers of Mennonites are the highest. The lack of information and awareness about the need to change practices, attitudes opposed to the acceptance of agri-environmental grants and a relative low level of education inhibit environmental program participation. Though again a site specific variable, religion can affect the degree to which people implement BMPs.

On the other hand, there are quite a few Old Order Mennonites who are very environmentally friendly and have become certified organic producers. However, for the most part the Mennonites participating in this study have small operations, low gross farm sales and lack the funding to implement BMPs even when so inclined.

The interviews conducted simultaneously with farmers within the CC sub-watershed of the Grand River revealed a much lower average level of education than E/S and a substantial lack of awareness about cost-share programs and even the EFP. This is unfortunate as a cost-benefit analysis of the CC study has shown that with even a small, uniform decline in net farm income of 3-5%, a sizeable reduction of sediment loading would occur if a zero tillage system were introduced while the net social benefits if implemented for the whole Grand River watershed could exceed $1.2 million (Filson *et al.* 2004).

The E/S area residents' high level of understanding of the need for and benefits from their environmental practices is probably a function of their much higher average educational level relative to the CC area. In CC many farmers do not feel they need to change their practices possibly because traditional practices are sacrosanct. Non-farmers on the other hand know there are problems in the CC area relating to agriculture though they, along with some of the farmers, insisted that the larger scale farms, or at least a combination of a few large scale farms and many small sized farms were usually the biggest sources of manure spills and other pollution problems, not many small sized farms, which is false (also see chapter 6). Agriculture is more intensive in CC than E/S and this is especially true of intensive livestock production so there is a greater need to adopt environmental BMPs in the CC than the E/S sub-watershed.

Over half of the farmers who understood that there are environmental problems in the area had adopted an EFP. The fact that the more environmentally conscious ones are the farmers who adopted these programs comes as no surprise. The concern here lies with those who have not implemented an EFP or other program, as they simply do not see any need to do so. The main advantage of the EFP relative to other plans is that it was developed by farmers for farmers. . While it has been well received by most Ontario's farmers, the problem has been that many farmers have completed only the initial stages of the plan (Smithers and Furman, 2003).

Because the RWQP, which provides funding for the adoption of BMPs, is tied to the adoption of the Canada-Ontario Environmental Farm Plan, some farmers may not participate in the RWQP because they have not done an EFP. On the other hand, many have completed an EFP because they knew they would not qualify for RWQP funding otherwise.

The NMA is meant to ensure farmers are managing their nutrients effectively. While many farmers in this study were not affected when interviewed, changes in the regulation will affect farmers if they were to expand their operations. With respect to regulations farmers in the study felt overall that mandatory regulation would be the only option if stricter environmental regulations were introduced, otherwise, farmers would not adopt. Gale *et al.* (1992) found that farmers agreed that voluntary approaches alone do not work very well and that mandatory regulation may be the best way to ensure farmers adopted conservation practices, however, a balance is required between mandatory and voluntary measures.

This chapter identified factors that inhibit farmers' adoption of BMPs that were not found in studies of the other southern Ontario watersheds: the lack of education and farmers whose religion affects every aspect of life. This underlines the fact that demographic factors impacting farmers' mindscapes with respect to how they interact with their landscapes, is site specific. In some cases, religion and the lack of formal education appear to be in contradiction with the need to introduce environmentally beneficial practices.

Farmers who had a higher level of education were generally more likely to have a better awareness of government programs than other farmers. Cultural aspects may also play a role in a farmer's decision to adopt. There are different groups of Mennonites and many subscribe to very strict Old Order beliefs, even though many also are quite indistinguishable from Protestant or other farmers in their acceptance of modern technology (Stevenson, 2003). Mennonite farmers generally felt that they are good stewards of the soil whereas many non-farmer participants did not feel that most Mennonite farmers in the area were good stewards. Many Old Order Mennonites' refusal to accept government grants for fear that it may put their pacifism at risk in the event of a war is an important factor which partially explains the low adoption of BMPs and participation in government-sponsored programs.

CONCLUSION

Several contradictions were assessed in this chapter including those that exist between farmers and non-farm rural people, between traditional farm practices and contemporary requirements, between numbers of livestock and the limited amount of land to spread the excess nutrients. Farmers in both sub-watersheds of the Grand were found to have a more instrumental view of the environment than non-farmers who expressed more concerned about protecting scenic qualities and biodiversity than farmers. The differences in perception are evidence of different mindscapes associated with being a farmer versus a rural non-farmer. While farm practices generally were perceived to positively affect residents' quality of life, some occasional farm practices such as the use of sewage sludge as fertilizer are objectionable to non-farm rural residents.

A number of factors affect both what motivates farmers to adopt conservation practices such as BMPs, as well as the barriers which may prevent them from adopting. Farm or other demographic characteristics such as farm income, education and religion play a significant role in these decisions. Overall, financial constraints are the most important factors affecting farmers' decision-making, as the lack of available funding available to farmers to make the necessary changes to their operations, is admitted to or complained about by many farmers. Legislation is of concern to farmers, despite the fact many have operations that are still too small to be regulated by the Nutrient Management Act. Many farmers feel that they have to look over their shoulders to make sure Big Brother is not watching them.

Non-farmers tend to believe that existing programs such as the RWQP and the EFP are beneficial tools to help farmers to understand and implement conservation measures on their farms. They recognize, however, the importance of incentives and consideration of social characteristics in the area, and whether programs or legislation would be successful. They also note, particularly in the Canagagigue Creek which does have such poor water quality, that it is important that players from a variety of sectors participate in mitigating the effects of pollution.

Whether a farmer will adopt a conservation practice has to be considered from many perspectives. This includes whether they trust the agency delivering the program or legislation, whether voluntary or mandatory, whether they see their peers are participating or not, whether they see the rural landscape in more of a multifunctional or unifunctional sense, farmers may or may not adopt BMPs or participate in an EMS. Because agriculture in Canada is generally not seen as multifunctional as it is in Europe, the positive externalities, such as wildlife preservation and land retirement are often not considered to be a benefit, regardless of whether funding may be provided to the farmer. Changing the concept of what is a benefit, and therefore introducing the idea of a multifunctional rural landscape, is important if environmentalism is to be adopted by the more reluctant farmers.

There are relationships among the farmers' demographic characteristics, farm structural variables, awareness, motivation and the degree of environmental program participation and BMP adoption. And while there is a widespread rural perception that a few large farm operations are more of an environmental threat than many small farms, the latter often provide a chronic threat especially when they under-invest in manure management though large farm spills can be more acute when they happen (Blackie and Tuninga, 2003).

Commercial BMPs are usually initiated first by large operators with the most income, farm sales and numbers of commodities produced. Those introducing the commercially viable BMPs tend to be more aware of the impact of agriculture on the environment and on people's perceived quality of life so they are more likely to participate in environmental programs like the EFP and RWQP and have a greater interest in adopting BMPs.

More non-farmers than farmers support mandatory nutrient management and the newly established agricultural regulations but of course it is the farmers who must implement nutrient management and BMPs and to do so they need society's help. On the other hand, most farmers would prefer voluntary programming and generally resent growing government intervention in agriculture because it interferes with their independence, creates more work for them and costs them scarce money. So the conflict between farmers and governments as well as between farmers and urbanites/exurbanites contributes to a contradiction between the nature of our present food production and the environment.

In: Agricultural and Environmental Security...   ISBN: 978-1-61668-156-2
Editor: Glen Filson   ©2011 Nova Science Publishers, Inc.

Chapter 5

FACTORS AFFECTING AGRI-ENVIRONMENTAL PRACTICE IN ONTARIO'S AUSABLE BAYFIELD WATERSHEDS

Paige Agnew and Glen Filson

INTRODUCTION

As indicated in chapter 4 above, public concern about water quality and environmental health in conjunction with a growing trend of agricultural intensification has fuelled the debate about which environmentally beneficial agricultural practices are appropriate. As a result, such Ontario regulations as the Nutrient Management Act (NMA) and the Clean Water Act (CWA) are being enforced to protect the environment. This reflects a gradual shift from the historical practice of farmer self-regulation (Montpetit, 2003) to one involving significantly more government regulation.

After reviewing the literature pertaining to landowners' adoption of environmentally beneficial management practices (BMPs) the extent of BMP participation of two Lake Huron drainage areas within the Ausable Bayfield watersheds, research with farmers within the Hobbs-McKenzie and Usborne areas, were undertaken. Landowners were interviewed and focus groups were conducted with them to determine their degree of BMP participation and their views about the environmental effects of farming. Social profiles of local residents in each drainage area were also generated using Agricultural

Statistics Canada data. Inferential and descriptive statistical techniques were used to determine the degree to which interviewees and those completing mailed surveys were representative of the residents.

RELEVANT LITERATURE

Many farm conservation measures are not profitable for farmers yet they are still desirable for the positive impact that they make on common property resources, land, air and water (Environics, 2006). Conservation stewardship extension advisors and conservation specialists should work with farmers of sites requiring remediation measures. This is always most effective when farmers join with environmental groups and conservation specialists in what Woodhill and Röling (1998) call *learning platforms*, as is the case with the Rural Water Quality Program (RWQP).

The Steering Committee for the RWQP contained representatives from 21 groups including at least 14 farmer organizations as well as federal, provincial and municipal governments and the Grand River Conservation Authority (GRCA) (Ryan, 2000). However, as Ryan (2000) has admitted, somewhat lower participation in the Environmental Farm Plan (EFP) process within some parts of the Waterloo Region limits the potential of the RWQP. On the other hand, the creation of the RWQP has increased EFP involvement.

Poor nutrient management on Ontario farms remains a particularly significant concern and the Nutrient Management Act (NMA) seeks to regulate this much more closely. Voluntary cost-share environmental programs such as the EFP and the RWQP run the risk of being implemented only by farmers who are already environmentally oriented given the cutbacks in extension staff which have happened since the Conservative Harris Government of the mid-1990s. Despite their lack of political palatability, the time may soon be coming to use downstream hydrological engineering, remote sensing and geographical information systems to identify farmers in greatest need of environmental remediation, so that public and private funds can be combined in the service of protecting the environment by building manure storages, riparian buffer strips and related best environmental management practices (Stonehouse, 1996).

This study was partly conducted to see if there was a relationship between landowners' adoption of BMPs and their quality of life as had been done by Filson (1996), Richmond *et al.* (2000) and Bucknell (2002). The first two of these studies found that people who earned more money tended to have better perceived quality of life and tended to take more pride in non-functional, aesthetic environmental amenities. While farm practices generally are perceived to positively affect residents' quality of life, some practices are objectionable and are the source of conflict between farmers and non-farmers such as the spreading of sewage sludge on farm land (see chapter 4). While the study by Filson (1996) indicated that farmers saw BMPs as desirable in protecting the aesthetic qualities of the environment (an indication of the multifunctional nature of Agriculture), the net economic cost associated with most of those initiatives were perceived to diminish quality of life by virtue of the labour and cost required to deliver them. For instance, manure management can detract from farmers' quality of life (labour, cost, odour) but can also protect environment, raise income and reduce conflict with neighbours. The implication of this is that farmers may be producing ecological goods and services (EGS) at their own detriment by adopting best management practices (BMPs). Thus, financial support is often a necessary incentive for farmers to adopt BMPs.

Studies by Napier and Brown,, 1993, Stonehouse, 1996; Filson, 2000 and McCallum, 2003 are additional examples of research that attempted to understand landowner program adoption by identifying trends and significant relationships between variables such as age, gender, education, farm size and farm types. Possible barriers to adoption as indicated within the former studies include additional economic strain, mistrust of government funding and programs, ineffective promotion of such programs, lack of skilled extension workers to assist in implementation and the perception that farmers are inherently 'good stewards' of the land already. Studies conducted by Filson (1993) and Filson and Friendship (1999) identified the motivating factors that lead to the adoption of such programs by rural landowners to be linked to factors related to quality of life. Environmental and scenic qualities as well as the physical conditions of their surrounding were perceived to play an integral role in providing a good quality of life (Richmond et al. (2000). Level of education, age, sex and religious affiliation are additional factors that could affect an individuals' degrees of concern for environmental issues (Van Liere and Dunlap, 1981; Arcury and Christianson, 1990; Howell and Laska, 1992, Stern et al, 1995) but the impact of these factors vary significantly as a function of context. The degree of concern for environmental issues partially predicts environmentally responsible behavior, such as adoption of BMPs (Fransson and Garling, 1999). Policy makers within government still have more to learn about the social and demographic factors that motivate rural landowners to be concerned about environmental degradation.

In 2001, the Maitland Valley Conservation Authority (MVCA) conducted a study of landowner motivations for the uptake of environmental BMPs. As indicated in chapter 4 above, the MVCA results indicated a relatively favourable attitude toward environmental conservation practices given the right circumstances. As one farmer told Kayak, (2001): "We would like to have our drinking water and our forests protected." Another said, "the regulations are strict but necessary for factory type farms. We need to protect ourselves from disasters!" Still another said, "the commodity prices [in 2001] are so bad you can't even buy barbed wire, so forget about fencing."

Blackie and Tuininga (2003) also contend that the majority of small livestock operations surveyed will require sizable financial investment to become compliant with Nutrient Management regulations; an investment that in the majority of cases they reviewed, is not financially feasible for the independent operator. Moreover, the issue of tile drain monitoring was highlighted within this study, as results indicate that only 42% of livestock producers were currently monitoring tile drain outlets. Blackie and Tuininga (2003) suggest that the adoption of this best practice would enable farmers to spot manure discharges in the tile drain and therefore provide an opportunity to respond to the spill prior to it entering surface water.

RESEARCH HYPOTHESES

Several hypotheses gleaned from our experience and the literature were created to guide the research process for this chapter. The hypotheses attempted to establish relationships between various independent and dependent variables in relation to the uptake of the Healthy Futures program[1] being operated at the time of the study among landowners. The validity of each of these statements was tested throughout the research process using a

---

[1] The Healthy Futures for Agriculture program was a $30-million program funded by the Province of Ontario for rural water quality projects including wellhead protection, plugging abandoned wells and implementing BMPs. Money was made available through the Conservation Authorities for doing such things as restricting livestock access to streams and creeks, for adopting nutrient management planning, etc. (Legislative Assembly of Ontario, June 3, 2003, Official Report of Debates (Hansard). Standing Committee on General Government, Nutrient Management Act, 2002. G-45.)

variety of qualitative and quantitative techniques discussed below. . These hypotheses suggested that :

- The level of education ought to be positively correlated with program uptake.
- A positive relationship between the age of the landowners and program uptake exists.
- Landowners in higher income brackets are more likely to participate in the Healthy Futures program.
- Landowners who had previously participated in similar Voluntary Environmental Programs (EVPs) were more likely to participate in the Healthy Futures program.
- There is a positive relationship between program uptake and the size of the landowner's farm..
- A positive relationship between the perceived quality of life of the landowner and the likeliness of program uptake exists.
- Landowner farm type/commodity produced are significantly related to program uptake.
- A positive relationship between the number of livestock units and program uptake exists.
- Landowner knowledge of the ABCA proximity to its location is significantly related to program uptake.

The lower Healthy Futures participation in Hobbs-McKenzie relative to Usborne can explained by the higher percentage of non-farm rural people in Usborne than Hobbs-McKenzie which in turn encourages Usborne landowners to participate more in stewardship programs.

METHOD

The two sub-watersheds targeted by this study are located in the Hobbs-McKenzie drainage area of the Ausable River and Usborne Township. A list of Hobbs-McKenzie landowners was provided by the Ausable Bayfield Conservation Authority, while the names of landowners in Usborne Township were located through the use of the Municipal Rolls in the South Huron Municipal Office. A 100% sample population in Hobbs-McKenzie (or 68 landowners) and 50% in Usborne or (150 landowners) were selected for use in this study. Each selected landowner was sent a letter in mid-November of 2003 explaining the nature of the study and requesting participation via telephone interviews.

Follow-up phone interviews were then conducted after the letters were sent out to increase the response rate but it was still initially inadequate. Ten interviews were conducted in the Hobbs-McKenzie area and ten were granted by Usborne landowners but most landowners contacted would not agree to participate in the study. An alternative course of action was designed in an attempt to secure a greater level of participation from targeted landowners. A survey package was created and mailed to 150 landowners, excluding those who had been previously interviewed or refused participation. Completed surveys were collected by house-to-house pick-up in 2004 in each of the targeted areas.

The names and contact information for each of the corresponding farm organizations (e.g. Ontario Federation of Agriculture {OFA}, Christian Farmers Association {CFFO}) within the targeted regions were documented as potential interviewees and/or additional entry points into the target communities. Initial contact was made with representatives from the farm organizations and the Rural Secretariat of the Ontario Ministry of Agriculture, Food and Rural Affairs (OMAFRA) during February of 2004. Interviews took place in the latter half of February and throughout the month of March.

A comprehensive social profile of the communities targeted by this study was completed using 2001 Agricultural Census Data and SPSS statistical software. The social profile helped determine the degree to which the views expressed and information provided by landowners are representative of residents in each area.

AGRICULTURAL CENSUS COMMUNITY PROFILES

Public concern for water and environmental health in conjunction with a growing trend of agricultural intensification has fuelled the debate in agricultural sectors worldwide as to what practices are deemed appropriate. This debate has evoked national political response in the form of new policy and regulation concerning environmental management, including the creation of BMPs and the NMA. The Canadian Government's stewardship driven approach has experienced varying levels of regional success.

Previous studies concerning landowner adoption of conservation practices have followed three explanatory lines of argument, one which focuses on farm and landowner characteristics as a view to understanding behavioral motivations. Data has been extracted from the Canadian Agricultural Census to construct detailed social profiles of the residents living within the Hobbs-McKenzie drainage area and Usborne Township for comparative purposes.

Figure 5.1 shows Western Ontario. To the map's west is Lake Huron. Usborne is in Huron Country and to its southwest, Hobbs-McKenzie is in Lambton County.

Figure 5.1. Usborne (in Huron County) and Hobbs-McKenzie (in Lambton County)

The boundaries of the Hobbs-McKenzie Drainage Area are delineated on a watershed basis and therefore do not follow the geographical boundaries as stipulated by Census Canada. The Hobbs McKenzie Drainage Area follows the natural course of the Ausable River, which traverses through more than one county and township. It was therefore not possible to glean specific community social profile information exclusively from Hobbs McKenzie landowners.

The residents within the Hobbs-McKenzie drainage area are mainly located in the Township of Warwick, within Lambton County. Therefore, the census data extracted was done by the census consolidated sub-division of Warwick Township, so that a general profile of this community could be created for comparison. Although some property owners within the Hobbs-McKenzie area are technically residents of the Municipality of Lambton Shores, Warwick Township with more "Hobbs-McKenzie" property owners was thought to be a more appropriate census area from which to build the community profile. Figure 5.2 shows the watershed map of Ausable Bayfield.

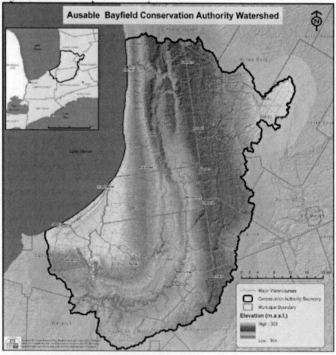

Source: (Tracy Boitson, ABCA, 2009)
Figure 5.2. Watershed Map of Ausable-Bayfield.

POPULATION CHARACTERISTICS

Usborne Township with 1535 residents has about one-third of the population of Warwick Township at 4,025 residents. There are a total of 445 people engaged within the agriculture sector in Usborne and 400 in the Township of Warwick. Overall, Usborne has a greater proportion of the total population in agriculture at 29% as compared to approximately 10% in Warwick. Although both areas are considered rural, almost 70% of the people in Usborne are non-farmers while almost 90% of residents in Hobbs-McKenzie are non-farmers. Table 5.1 below provides a numerical comparison.

Table 5.1: Population Breakdown by Number of Agricultural Operators

Characteristic	Warwick				Usborne			
	Number		%		No		%	
Total Population	4025		100		1535		100	
Agricultural Operators	400		10.0		445		29.0	
Operators by Gender (M/F)	285	115	71.0	29.0	350	95	78.7	21.3

Usborne and Hobbs-McKenzie are quite similar in terms of age structure. The average age of farm operators in Warwick is 47.6 years of age and 50.8 years of age in Usborne.

Table 5.2: Community Breakdown by Census Age Category

Age Category	Hobbs-McKenzie	Usborne
Under 35	60 (15%)	40 (9%)
35-54	220 (56%)	245 (54%)
55 +	115 (29%)	165 (37%)
Total	395 (100%)	450 (100%)

The census further breaks the age data into three categories (shown above). In both areas, just over 50% of the population is between the ages of 35-54 years of age and approximately 30% are 55 years of age or more. From this information it can seen that both Warwick and Usborne have a large proportion of retired residents in relation to total population size. With a relatively small proportion of the population under 35 years of age (Warwick 15.0% and Usborne 9.0%) both areas can be characterized as having an aging population.[2]

FARM CHARACTERISTICS

Table 5.3: Farm Sizes in Warwick and Usborne

Size Category (in acres)	Warwick (# of Farms Reporting)	Usborne (# of Farms Reporting)
10 to 69	16	47
70 to 129	36	70
130 to 179	55	36
180 to 239	38	49
240 to 399	34	48
400 to 559	44	26
560 to 759	29	21
760 to 1,119	9	11
1,120 to 1,599	8	7
1,600 to 2,239	2	1
Total	271	316

[2] The totals indicated in Table 5.2 are greater than the census data total number of farmers in Usborne Township. This difference can be attributed to rounding errors in the census raw data.

Information regarding farm size provides an indication of what type and size of operations exist within a community. More specifically it indicates whether operations are small, medium or large intensive operations. The agriculture census classifies farms in Warwick and Usborne by size in acres (Table 5.3).

The table above summarizes information by number of farms reporting.[3] Figure 5.3, below, shows the farm sizes for Warwick and Usborne in acres.

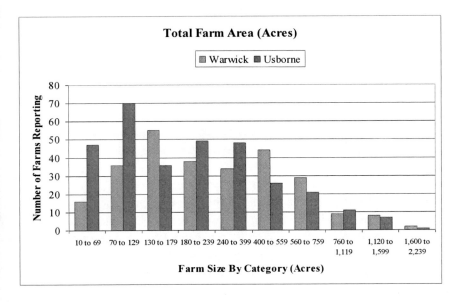

Figure 5.3: Farm Size by Category in Acres

The figure above shows a side-by-side comparison of farms within the two areas of study. Usborne has a much higher percentage of farms in the 10-69 acres category compared to Warwick. Warwick however, has a greater number of farms in the 70 to 129 acre category. In the 130 to 179 acre category Warwick has a greater percentage of total farms than Usborne at 21% to 11% respectively. Both communities possess a comparatively similar cross-section of farms by size category, primarily small to medium size operations. Warwick Township has a greater number of recreational or lifestyle farms,

---

[3] This is important to note because once again the totals indicated for both areas are smaller than the total number of farms indicated at the beginning of this report and therefore the information provided by the census is somewhat incomplete.

indicated in the first size category. The majority of farms in both target areas are between 10-400 acres, categorized as small to medium sized operations.

There was a slightly higher percentage of Usborne than Warwick farms that had no off-farm work although Usborne had a greater number of respondents in the category 'more than 40 hours'. It should be noted, however, that most farmers in both areas had no recourse to off-farm employment.

BACKGROUND INFORMATION ABOUT THE SAMPLE POPULATIONS

Hobbs-McKenzie

Of the 28 interviews and questionnaires completed in the Hobbs-McKenzie sub-population, 27 respondents were male and one was female. Approximately 82% or (23) of respondents were farmers and 18% or (five) were non-farmers. There was a mixture of small to medium farm operators, with an average of 31.5 years of farming experience among respondents. While 21 respondents indicated that they were born in Canada, seven said that they immigrated to Canada from either the Netherlands or Belgium and have been in Canada for an average of 42 years. Among the farmers interviewed, there was a fairly even mix of farming types represented, with field crop farming reported by the greatest number of respondents at 21% or (5 out of 23).

Table 5.4: 2002 Gross Farms Sales by Respondent

Sales Category	Hobbs-McKenzie (%)	Usborne (%)
Less than $49,999	3 (33.3)	8 (32)
$50,000-$79,999	0 (0)	2 (8)
$80,000-$119,000	0 (0)	1 (4)
$120,000-$159,999	1 (11.1)	4 (16)
$160,000 and Over	5 (55.5)	10 (40)

Among the farmers interviewed, who provided the information, a range of approximate gross farm sales were obtained in 2002.   Out of 23 farmer respondents in Hobbs-McKenzie, for instance, only nine provided this information compared with 25 out of 40 from Usborne (see Table 5.4).

Of the 40 questionnaires completed by landowners in Usborne Township, 35 were farmers and five respondents were non-farmers. Of the total questionnaires received 75% (30) were from men and the remaining 25% (10) were from women.

Table 5.5: Study Participants by Year of Birth

Date Born	HM Respondents	Usborne Respondents
Before 1930	3	3
Before 1940	2	5
Before 1950	11	14
Before 1960	7	14
Before 1970	5	4
Total	28	40

Table 5.5 shows Study Participants by Year of Birth. Respondents' ages varied from 35 to 80+ years of age in both sample populations with approximately 65% of the population born in the 1940s and 1950 (in their 50s and 60s). Thirty-nine Usborne participants were born in Canada and one respondent had immigrated to Canada from Belgium 52 years ago. The majority of study respondents from Hobbs-McKenzie and Usborne said they had completed high school and some (35% and 32% respectively) had completed post-secondary training.

Table 5.6 shows the level of education of participants from Usborne and Hobbs-McKenzie sub-watersheds.   Overall, there were no significant differences in the age structure or educational background of the communities targeted by this study. A review of relevant literature suggests that in many previous studies, landowner's age and level of education have been significant variables in the determination of program uptake. This was not true for this sample though this issue is reviewed below under Statistical Analysis and Interpretation.

Table 5.6: Participants' Level of Education

Level of Education	Hobbs-McKenzie Residents	Usborne Residents
Primary School or Less	4%	5%
High school	61%	63%
Post Secondary	35%	32%
Total	100 %	100 %

The farmers interviewed produced a cross section of commodities, including cash crops and livestock. Table 5.7 below shows a breakdown of farm respondents by farm type or commodity produced. Further analysis reveals 40% of respondents in both target areas had livestock while a slightly higher percentage of respondents from Usborne engaged in field cropping at 37.5% compared to 21.7% in Hobbs-McKenzie. No respondents from Usborne reported involvement in tender fruit production compared to the two or almost 10% of total respondents from Hobbs-McKenzie.

Table 5.7:   Participants' Commodity Types for Hobbs-McKenzie and Usborne

Farm Type by Commodity							
	Field Crop (%)	Beef/ Dairy (%)	Poultry (%)	Swine (%)	Forages (%)	Vege-tables (%)	Fruit (%)
Hobbs-McKenzie	5 (21.7)	4 (17.4)	2 (8.7)	3 (13.0)	4 (17.4)	3 (13.0)	2 (8.7)
Usborne	15 (37.5)	9 (22.5)	1 (2.5)	6 (15.0)	7 (17.5)	2 (5.0)	0 (0)

Findings

Program Adoption

It must be noted that while 28 questionnaires were completed in Hobbs-McKenzie and 40 in Usborne, some respondents chose to omit certain answers. This may account for the discrepancies in some data present within the following tables. More landowners in Usborne Township interviewed indicated having knowledge of incentive funding for implementation of BMPs, and an overview of 2001-2003 data and information collected through this research indicates that Usborne landowners had a much higher rate of participation in past environmental farm programs (CURB, SWEEP, LSP) and in the 2003 Healthy Futures program. The response rate for each instrument is provided in table 5.8 below.

Table 5.8: Response Rates for Each Data Collection Instrument

Data Collection Techniques Completed					
Sample Populations	# of Land-owners Sampled	Phone	Mailout Survey	Focus Group	Response Rate
Hobbs-McKenzie	68	8	20 **	4	47%
Usborne Township	150	10	30 **	6	31%

**Please note that focus group participants completed the research questionnaire.

Of the 38% of Usborne respondents who indicated they had received Healthy Futures funding, the following projects were completed: clean water diversion, septic system upgrades, fertilizer wash pad and windbreak by tree planting. The 18% of Hobbs-McKenzie who said that they had received Healthy Futures funding for BMP implementation indicated completing the following projects: nutrient management plan assistance, cattle fencing (livestock restriction) and clean water diversion. These trends of landowner participation are reinforced upon reviewing the ABCA Grant Program Summary of 2003 found in Table 5.10 below.

Table 5.9: Respondents' Participation and Awareness of Stewardship Programs

	Participation in Healthy Futures		Awareness of Current Programs		Participation in Environmental Farm Programs		Interest in est. Environmental Goals for Property		Completed EFP		Completed NMP	
	Yes (%)	No (%)	Yes (%)	No (%)	Yes (%)	No (%)	Yes (%)	No (%)	Yes (%) (%)	No (%)	Yes (%)	No (%)
Hobbs-McKenzie	18	82	18	82	14	86	54	46	36	64	18	82
Usborne	38	62	13	87	20	80	75	25	35	65	12.5	87.5

Table 5.9 presents the respondents' degree of participation in and awareness of stewardship programs.

Table 5.10: Healthy Futures Projects Completed by Study Area

Municipality	Healthy Futures 2001-2003	
	Projects Completed	Total Grant Dollars
South Huron	88 (48 in Usborne)	$237,549
Warwick (including Hobbs-McKenzie)	3	$3,306

The Head of the Stewardship Department for the ABCA explained how the ABCA works with farmers in the Ausable Bayfield watersheds.

"The farmer will come to us and he may know what he wants to do, such as a fence 1000 meters long and we help him with his application for the money. Sometimes there is run-off from his yard such as from corn sileage. We provide technical assistance to help him know what will work on that farm. We have access to information through OMAFRA and other government agencies" (interview with Monk, June 15, 2006, ABCA).

She observed that "farmers would like to have annual payments, for example, for creating riparian strips that takes land out of production." She noted that the ABCA does not have annual grants for environmental projects such as the Grand River Conservation Authority has but the ABCA does help with the capital costs. She noted that the only thing that blankets the entire province is the EFP but the environmental programs "come and go" making them problematic from the farmers' point-of-view. Key to the relationship between the CAs and the landowners is the rapport that can be established between them. She commented that

"It is about relationships and developing a rapport with the farmers. For example, one farmer I've been working with and his cattle were loose and went through a kilometer of the stream. He phoned the ABCA's General Manager and he said the Head of Stewardship will come out and see you. The farmer was afraid I (the Stewardship Head) would tattle to the Federal government about him." [The Stewardship Head commented about his cattle, pointing out that she had grown up on a farm and liked working with cattle. "Then he started with that project and he's gone on to do others. Projects have come and gone and we find that frustrating and so do the farmers" (interview, June 15, 2006).

Table 5.11: Barriers to Greater Uptake of BMPs in Hobbs-McKenzie Sub-watershed

Barriers to Adopting Best Management Practices
Lack of Clarity and Educational Constraints: Insufficient awareness of impact of farming practices outside of own operation (e.g. notion of being part of a greater ecosystem, cumulative effects). Of the possible environmental concerns in community. In the design and delivery of current programs Why are these programs relevant/important? What agri-environmental programs and funding are available? Who to contact? How to gain access to such programs? What are the standards/regulations that must be met in program?
Mistrust/Animosity Perceived untrustworthy nature of government/conservation authorities Perception that programs aren't in best interest of farmers, instead yielding to the government's agenda. Dislike of current level of government regulation/involvement in farm operations.
Perceived Impact on Economic Viability of FarmingOperation Cost-share notion not received favorably. Extra costs involved in times of diminished farm revenue and low commodity prices. Associated costs with new NMA regulation. Weak connection between program investment and visible economic /operational improvements to farm.
Perceived Quality of Current Practice Common attitude: current practices are already 'best practices'.

The top reasons indicated by survey participants for non-adoption of voluntary stewardship program in both target areas include: 1) unaware of such initiatives 2) have no environmental problems 3) extra costs make it unfeasible no funds remaining 4) application rejected 5) low financial return on investment 6) wary of government agencies and programs.

While a few livestock producers interviewed indicated that they had a Nutrient Management Plan (NMP) in both areas, most admitted that their NMPs were not in compliance with local Nutrient Management by-laws. It must also be noted that only farmers wanting to expand their livestock operations plus those who have 300 animal units or more are required to create a NMP.   Also, in Usborne, 14 of 40 or 35% of respondents indicated that they had completed an EFP and 10 of 28 or 36% reported having done so in Hobbs-McKenzie.

Of those respondents from Hobbs-McKenzie and Usborne, only a small percentage were aware of the stewardship grant funding that is available in their areas, at 18% and 13% respectively. The low percentage of landowner awareness however, is not reflected in the much higher percentage of program uptake in Usborne compared to other adjacent municipalities.

A problem with communication between the ABCA and local landowners can be inferred from this which was corroborated by the questionnaire results. Only 29% of respondents in Hobbs-McKenzie said that they would contact the ABCA for information on stewardship programs compared to 88% of respondents in Usborne. OMAFRA was mentioned by 18% of Hobbs-McKenzie residents as their contact organization of choice, with the remaining 53% of respondents unsure of who to contact regarding stewardship programs. Again, this draws attention to the visibility of the ABCA within each target community and the presence that is maintained by the ABCA in the eyes of Usborne residents.   It may also reflect the feelings encountered in our focus groups and interviews of animosity and mistrust toward government agencies intimated in the Hobbs-McKenzie focus group.

All interviewees questioned about the development of environmentally sustainable farms indicated this was an important personal and community objective.   Interview respondents in both target communities characterized environmental stewardship as a moral and social obligation. However, when asked about environmental concerns in their community, drainage issues, water quality and proper manure spreading were the only issues identified.

Government Regulation

Several respondents in both target areas declined to answer questions about government regulations. Of the 24 respondents from Usborne who did answer, 15 felt that there is currently enough regulation, five felt that there is not enough and four indicated that too much regulation exists.

**Table 5.12: Participants' Views about the Appropriate Level
of Government**

Level of Regulation	Hobbs-McKenzie	Usborne
Not Enough	11%	22%
Enough	35%	63%
Too Much	54%	15%
Total	100%	100%

In Hobbs-McKenzie, out of the 26 questionnaire responses, 15 felt that there is too much regulation, eight said there is enough and three replied that there is not enough. This data is found in Table 5.12.

Regulation in Agriculture

Just over half or 54% of the Hobbs-McKenzie respondents felt that there is currently 'too much' government regulation in agriculture whereas only 15% of Usborne respondents felt that is the case. In Usborne however, most respondents (63%) indicated that there is currently 'enough' agricultural regulation by government.   One landowner commented that "These are all good initiatives but generally policy and regulation would appear like they are prepared by a non-agricultural sector and thus they tend to fail."

On the other hand, 22% of respondents from Usborne actually thought that there is not enough government regulation and said that greater enforcement of current regulations is required on behalf of government agencies. In the data samples provided by the two target communities, there are a greater number of non-farmer respondents from Usborne compared to Hobbs-McKenzie, which may account for these differences.

The Head of Stewardship for the ABCA commented that "We're trying to make regulations for everyone but only 95% do them and there are some people who will do them only because there are regulations and others who will only do them because there are no regulations." (interview, June 15, 2006).

These results agree with those of chapter four for two sub-watersheds of the Grand River which found that non-farmers tended to favour government regulation of agriculture, while farmers generally preferred little or no government intervention in their operations. Moreover, several non-farmers felt that there are insufficient consequences or penalties for farmers in violation of environmental standards, by-laws or regulations.

By contrast, in the focus group one Hobbs-McKenzie farmer said "Everybody's criticizing agriculture. No one's talking about septic systems or industrial waste." However, another farmer thought that some inroads we're being made with septic systems. The first farmer angrily responded "No! We're being blamed for everything!"

When questioned about government incentives and uptake of stewardship programs, most respondents in Hobbs-McKenzie and Usborne indicated that greater subsidization and long-term program funding would increase adoption of stewardship programs (Table 5.13).

Table 5.13: Respondents' Preferred Government Incentives

Preferred Government Incentives	Hobbs-McKenzie (percent)	Usborne (percent)
Assurance actions meet regulations	3 (11.5%)	12 (33.3%)
Greater subsidization (more funding)	17 (65.4%)	20 (60.6%)
Increased clarity of regulations	6 (23.1%)	2 (5.6%)
Other (Tax Incentives)	0 (0%)	2 (5.6%)
Total	26 (100%)	36 (100%)

As well, public awareness campaigns and educational initiatives were viewed by some respondents in both targeted areas as crucial in the process of defining local environmental concerns, highlighting stewardship program funding available and clarifying the regulatory standards of new agri-environmental legislation. Comments received from farmers and non-farmers alike in Hobbs-McKenzie suggested that if programs were streamlined to focus on farmer needs and made sense from a farming perspective, more farmers would be inclined to adopt such programs.

Perceived Quality of Life

Most farmers in both communities did not discern a noticeable difference in their quality of life from 5 years ago until the present. While farming respondents were more concerned with the diminished quality of life from an economic perspective, non-farming landowners focused on the diminishing quality of the physical environment including reduced air and water quality. Non-farm or small farm operators in general reported little change in their quality of life from 1998 to 2003 with the latter indicating that improved supplemental off-farm income has increased their financial security. Respondents' perceived (QoL) reported for 1998 positively correlated with that reported from 2003 ($\sigma=0.573$; N=68; P$\leq$0.000). Respondents did not experience a significant change in their quality of life over that five year period.

STATISTICAL ANALYSIS AND INTERPRETATION

The analysis in this section was conducted with the questionnaires from a survey of 210 respondents. There were 271 farmers in Hobbs-McKenzie so a sample providing a sampling error of 10% was selected equal to 60 respondents. In Usborne, 150 out of 316 landowners were selected providing a sampling error of about 5%. Initially only 10 landowners in Usborne (U) and 8 in Hobbs-McKenzie (HM) agreed to be interviewed by phone. These farmers were subtracted from the two samples leaving 58 from HM and 140 from U. A survey package was created and mailed to the remaining 198. A focus group was held with 4 additional farmers in HM and an additional 6 in U. The final sample of 28 respondents therefore had a sampling error of about 15%. Nonetheless, the results of this study are broadly similar to those of the other southern Ontario watershed case study regions discussed in chapters 4 and 6.

It was difficult to gain participation due to both the length of the questionnaire and the fact that in the immediate aftermath of the nearby Walkerton tragedy, it was very difficult to obtain agreement from farmers to answer questions about their adoption of BMPs. It is also true that many people consider the funding agency, the ABCA, to be an arm of government, despite the fact that conservation authorities are actually at arms length from government. There is a strong anti-government feeling among farmers expressed directly in the many rural signs such as "Private Property! Back Off Government!" and "Farmers Feed Cities!"

This notwithstanding, the quantitative analysis of the data enabled us to identify general observations regarding landowner behaviours, and mindscapes as well as the trend regarding program uptake. Triangulation of information gleaned from the Agricultural Census Data, key informant interviews and qualitative observations extracted from the field increases the validity of study results to follow, yet the sampling error reduces the reliability of the results somewhat.

A cross tabulation on data suggests there is no significant relationship was found between education and program uptake ($\chi^2=0.094$, N=62, P$\leq$0.954). In the case of income, the following tables (Table 5.14) show a numerical breakdown of landowner Healthy Futures participation cross tabulated with their reported gross farm sales from what they reported from 2002.

Table 5.14: Cross Tabulation Farm Sales and Healthy Futures Uptake

		Have you taken advantage of Healthy Futures funding for BMPs?		Total (%)
		Yes (%)	No (%)	
Category of farm sales	less than $119,999	2(10.5%)	8 (18.6%)	10 (16.1%)
	more than $120,000	10 (52.6%)	9 (20.9%)	19 (30.6%)
	no answer	7 (36.8%)	26 (60.5%)	33 (53.2%)
Total		19 (100%)	43 (100%)	62 (100%)

There is a significant association at the 0.05 level of significance between gross farm sales and taking advantage of Healthy Futures funding for BMPs ($\chi^2=6.24$; N=62; P$\leq$0.044). Those with gross sales over $120,000 were relatively more likely to take advantage of the program than those with lower gross sales.

It was hypothesized that a relationship would exist between landowner farm size and willingness to adopt BMPs in the target sub-populations. The data indicates a significant association at the 0.05 level of significance between farm size and taking advantage of Healthy Futures funding for (BMPs) (χ^2=8.51; N=62; P≤0.014) with relatively larger farms having tended to take advantage of the program more often than smaller operations. There is also a significant correlation between farm size and having taken advantage of Healthy Futures funding (σ=2.885; N=62; P≤0.005).

It was hypothesized that there would be a correlation between farm type and landowner uptake of the Healthy Futures program, with livestock owners being less likely to participate. However, there was a positive correlation at the 0.01 level of significance between the variables dairy production and uptake of the Health Futures Program (σ=0.317; N=5; P≤0.012). There was also a positive correlation between the production of forages and participation in the Healthy Futures program and a positive Spearman correlation (σ=0.322; N=9; P≤0.011). The small sample sizes suggest that these findings should be considered with caution.

It was predicted that farmers nearing retirement would be less willing to commit to additional voluntary expenditures on their property. A cross tabulation was performed on these two variables to test this statement. The six age categories indicated on the questionnaire were collapsed into two main categories for analysis in an attempt to secure greater levels of statistical significance. The hypothesis was rejected as there was no significant relation between these two variables. There was no association between previous participation in Environmental Voluntary Programs (EVPs) and uptake of the Healthy Future program and no significant relationship were found between the variable quality of life and uptake of the Healthy Futures program.

It was hypothesized that a landowner's awareness and preference of the ABCA for obtaining information on EVPs would positively correlate with program uptake. Respondents were asked to indicate what organization they would contact for information regarding stewardship programs in their area. Three general responses were given, including ABCA, OFA and OMAFRA. A cross tabulation was run on these two variables and some significant results emerged (Table 5.15).

Table 5.15: Cross Tabulation of Preferred Contact Organization and Healthy

		Have you taken advantage of Healthy Futures funding for BMPs?		Total (%)
		Yes (%)	No (%)	
What organization would you contact for information regarding stewardship programs?	ABCA	14 (73.7%)	22 (51.2%)	36 (58.1%)
	OMAFRA	3 (15.8%)	3 (7.0%)	6 (9.7%)
	OFA Rep	1 (5.3%)	4 (9.3%)	5 (8.1%)
	Don't Know	1 (5.3%)	14 (32.6%)	15 (24.2%)
Total		19 (100%)	43 (100%)	62 (100%)

Futures Uptake

There was no significant difference at the 0.05 level of significance between landowner preference for contacting the ABCA and uptake of the Healthy Futures program There was a positive relationship between whether or not respondents had done an EFP and the size of their acreage (χ^2=13.45; N=63; P≤0.001). The number of livestock units was also correlated with whether or not the farmers had done an EFP (σ=0.418; N=63; P≤0.001), thus those with more livestock were also more likely to have done an EFP. A farmer commented that

"The $1500 that you get from EFP [it's now possible to obtain as much as $30,000] you get only after putting down $5000. It sounds like a lot of money but it doesn't do anything. If they want to get serious with helping they should give interest free loans. We look after all these things ourselves. If we got what we deserve from the market, then it would be fine."

The expectation of whether or not the NMA would impact respondents' farms was positively correlated with whether or not the farmers had ever implemented an environmental farm program (σ=0.425; N=59; P$\leq$0.001). Whether or not people had a Nutrient Management Plan was positively correlated with the number of livestock units (σ=0.365; N=63; P$\leq$0.003) but negatively correlated with the degree to which their income arose from farming alone (σ=-0.360; N=37; P$\leq$0.029).

An inverse correlation was found between participant's assessment of the degree of regulation in agriculture with whether they had completed a Nutrient Management Plan (σ=-0.252; N=62; P$\leq$0.050).

The result of the data analysis when viewed in conjunction with additional analysis gleaned from the other methods of data collection (triangulation), provide an understanding of the factors and variables linked with the uptake of EVPs in rural communities. Clearly those with sales over $120,000 were most likely to participate in the Healthy Futures program. An association and correlation was also found between farm size and program uptake, as participants with relatively larger farms most often took advantage of the Healthy Futures program which, as with the relationship with GFS, is consistent with the findings reported in chapter 6. Thus, farm operators with more land and subsequently greater financial stability more often took advantage of incentive funding for the implementation of BMPs.

Observations recorded during focus groups and personal contact with landowners corroborates these results. One farmer said he was having a hard enough time making a living without having to implement excessive regulations. Farmers with smaller, more vulnerable operations from a logical standpoint would be less inclined to put forth their minimal resources to participate in something that is voluntary, especially if the outcome is difficult to measure from an operational standpoint. From an administrative standpoint, this possibly underscores a need to re-evaluate the levels of financial assistance available to farmers with smaller operations and modify or streamline future programs that target this category of operators.

The majority of farm commodities when cross tabulated with the variable of Healthy Futures uptake yielded no significant relationships. There was however a strong association between dairy producers, forage producers and program uptake. In many cases, each of the respondents who indicated participating in these categories of farming also indicated average (GFS) of $75,000 or more per year. Dairy and forage producing respondents also reported a farm size of 200 acres or more on average, reinforcing the notion that larger, more prosperous farmers were most inclined to participate in Healthy Futures. It should however be noted that there were some respondents with large farms and sizeable farm sales who declined to participate in the Healthy Futures program, so this relationship cannot be read off from statistics alone.

A positive correlation was found between the number of livestock units and respondents having a Nutrient Management Plan (NMP), as those with a greater number of livestock units were more likely to have completed a plan. This result corresponds with previous observations in that beef producers had a greater tendency to participate in Healthy Futures than producers of other commodities. Those respondents with larger farms were also more likely to have completed an EFP and there is a correlation between those who had completed an EFP and taken advantage of Healthy Futures funding.

Initially it was hypothesized that a diminished quality of life in rural areas, particularly in Hobbs-McKenzie, had contributed to poor uptake of the Healthy Futures program. Results indicated that the majority of respondents did not discern a noticeable difference in their quality of life between 1998 and 2003. Many respondents did however indicate that an increase in off-farm income has been critical in maintaining quality of life in the face of market vulnerability and financial instability of private farm operations. A few respondents from Hobbs-McKenzie wrote lengthy answers to this question, detailing the hardships and grievances felt by their family and local farming community and more specifically a lack of funds for operational daily maintenance let alone additional program expenditures. One said "our quality of life is going to hell in a hand basket."

Observations recorded during focus groups echoed these sentiments in both target areas, with participants in Usborne and Hobbs-McKenzie listing greater financial assistance as the second largest motivator for future program participation. Based upon the results recorded, increased financial assistance will play a critical role in determining degree of participation from the targeted communities.

Focus Groups

As a third method of data collection, focus groups were organized in each of the target populations. The local representatives of the Ontario Federation of Agriculture (OFA) were contacted to encourage landowner participation in each area. Each focus group lasted approximately two hours and consisted of group roundtable discussion. Three main questions were posed to landowners for discussion during the focus group and a ranking exercised was completed at the end of each session.

The questions included:

- What is the extent of agri-environmental programs within this area?
- Why would there be more uptake of BMPs in one watershed of the Ausable River than in another (e.g. farm or commodity type, farm size, gross sales, landowner's education level etc)?
- What motivates farmers to adopt BMPs (e.g. technical, financial or social responsibility)? Rank what you consider to be the top three motivators for programs uptake.

The first focus group was held at the Ausable Bayfield Conservation Authority on April 7, 2004 for landowners within Usborne Township. There were six participants at this focus group.

Respondents felt that there was favorable uptake of agri-environmental programs in Usborne Township specifically and the county of South Huron in general. Several agri-environmental programs were discussed by participants including Private Land Stewardship, Land Stewardship II, Great Lakes Renewal, Clean Up Rural Beaches (CURB) and Healthy Futures. All six respondents had completed an EFP. In addition, participants listed several BMPs that they had either personally implemented on their property or that they could report having seen implemented on adjacent properties. When asked about CURB a farmer said "CURB? Oh yes, we were in that. We did eves troughing, milkhouse wastes; we put in $15,000 out of a $50,000. There was $10 for manure runoff and the rest for milkhouse wastes." The following is a list of projects participants indicated receiving Healthy Futures grant funding for:

1. Windbreaks and fragile land retirement
2. Septic tank upgrades
3. Well upgrades and decommissioning
4. Water diversion
5. Livestock access restriction from watercourses
6. Roofing over cattle holding pen
7. Milk-house treatment

In response to the question of regional variation in BMP uptake, participants indicated a number of possible explanations. Participants felt that larger farm operators were less likely to participate in agri-environmental programs as they tend to be busier and more concerned with volume of production over quality. (In fact our statistical analysis revealed the opposite). Farm ownership was indicated as a variable involved with program uptake, as renters are less likely to make financial investments in land that they do not own. The economic viability of a farming operation was indicated by focus group participants to be extremely important in determining level of participation in such voluntary programs.

The Healthy Futures program requires landowners to supply the capital costs of the project up front with compensation to follow upon project completion. Participants felt that for many 'economically struggling' farmers these programs are cost prohibitive. Usborne participants also indicated that the 'closeness' or 'tight-knit' nature of their community and the close proximity of the ABCA has been an asset in disseminating information about voluntary stewardship programs and encouraging greater program uptake.

When questioned regarding motivation for the uptake of agri-environmental programs, participants contributed a long list of factors. These were then condensed during the ranking exercise, when participants were asked to choose the top three motivators for program uptake. The following are the top three motivators for program uptake as indicated by Usborne focus group respondents.

Legislation
2.a) Available funding*
2.b) Belief that it is the right thing to do*
3.   Cost effectiveness of initiative to farming operation
     *tied during the ranking process

The second focus group was held with Hobbs-McKenzie residents at the St. Williboard Credit Union on April 14, 2004. There were four participants present at this focus group and the exercise lasted approximately two hours. Participants were asked the same three questions as those posed in the Usborne focus group and completed a similar ranking exercise.

Hobbs-McKenzie participants had somewhat different views concerning voluntary stewardship programs than those provided by Usborne residents, but there was some commonality of response. Hobbs-McKenzie participants indicated that some local residents have implemented BMPs, but that the cost prohibitive nature of such programs and a lack of sufficient information regarding such programs has affected program uptake.

While respondents had some knowledge of such stewardship including CURB and Healthy Futures, fewer completed BMPs were reported by the participants of the Hobbs-McKenzie focus group.

These included:

1. Wind breaks (tree planting)
2. Cattle fencing (application for funding was denied)
3. Water diversion.

When questioned about the varying level of program uptake between communities, Hobbs-McKenzie participants indicated that growing animosity and distrust of government agencies resulted in limited local participation in stewardship programs. Also the funding application processes were perceived to be too lengthy and cumbersome. One farmer said "The CURB program made me so mad. Part of the problem is the bureaucracy. The people on the Board were not even agriculturalists. They didn't know what they were doing."

The fact that federal Department of Fisheries and Oceans (DFO) inspectors had come into the area and may end up charge a local landowner with failure to comply with environmental regulations was particularly galling for a couple of members of the focus group.

Later the Head of Stewardship at the ABCA observed that Environment Canada had decided a few years ago to be more proactive in enforcing the Fisheries Act. Farmers were told that they had one year to get their cattle out of various streams and creeks. Particularly upsetting to the farmers was the fact that Environment Canada would drive into their farm lanes unannounced (interview with Kate Monk, June 15, 2006).

Some participants also indicated that the literature distributed about stewardship programs is unclear to the average person, making it difficult to understand what funding is available for different types of projects. Participants also felt that the ABCA's lack of visibility within their community and physical distance to their office in Exeter also contributes to landowner's non-adoption of stewardship programs. Some landowners also complained of the difficulties of applying for grant funding, given the fact that their property was located in adjacent counties. Varying amounts of funding and different application processes by county was a source of frustration for Hobbs-McKenzie landowners and additional disincentive for program uptake.

Hobbs-McKenzie participants placed emphasis upon the need for the ABCA to secure a greater role for local landowner participation in program design and delivery, saying that these programs appear to be created, as one farmer said, echoing another mentioned above, "by people with no farming background and understanding of the farmers' needs and challenges". Group discussion first produced a list of 'disincentives' for program adoption, which then led into a discussion of what factors would motivate them as landowners to participate in voluntary stewardship programs. The list of disincentives put forth by Hobbs-McKenzie landowners included:

- Process too lengthy and cumbersome (too much bureaucracy)
- Farmers feel threatened by 'negative' image of agriculture
- Mistrust of government
- Aging farming community
- Lack of money for project capital costs.

There is some degree of commonality between the focus groups' responses regarding landowner motivation, specifically the correlation between the economic viability of the farming operation and farmer's willingness to invest money in a voluntary stewardship program. All focus group respondents agreed that in times of economic uncertainty, as is the case currently for some farmers, farmers simply do not have extra money to invest in projects that they perceive to produce a low return on their investment. One person suggested that there is an order in which farmers spend: they first spend their money making things, secondly they spend it on their homes and thirdly they spend their money on the environment.

The list of motivations put forth by Hobbs-McKenzie focus group participants are as follows:

- Greater involvement of farm organizations (farmers) in process, equal decision-making between farmers/non-farmers about program guidelines and processes.
- Greater level of financial compensation. Long-term programs with greater clarity of expectations and guidelines for the landowner.
- Straight-forward application process.
- Assurance of confidentiality for information disclosed through program participation.

A discussion of the findings with ABCA officials revealed that the Hobbs-McKenzie farmers may have felt unfairly picked on by local and even federal government officials.

DISCUSSION

When referenced with other pertinent topical literature our observations gain credence. Studies by Fransson and Garling (1999), Featherstone and Goodwin (1993), Lasley *et al.* (1990), Rahm and Huffman (1984), Christensen and Norris (1983), as well as Van Liere and Dunlap (1981) all contend that variables such as age, farm income, off-farm income, farm size and farming experience can, under certain circumstances, influence farmer's mindscapes and attitudes towards environmental conservation. For example, in a 1990s study of southwestern Ontario farmers' adoption of the SWEEP program, Serman and Filson (1999: 69) found "...that the number of crops cultivated, the size of the farm operated, the level of gross sales and the level of education all had the most significant influence on the farmers' adoption index which was comprised of the number of soil and water conservation practices adopted."

Initially the age structure of the communities targeted was thought to account indirectly for the varying uptake levels of BMP programs in the target areas. The formulation of this hypothesis was informed by Napier and Brown (1993) and Gould *et al*, (1989) who suggest that younger less experienced farmers are usually more inclined to accept the merits of environmental stewardship initiatives. The foundation of our initial hypothesis was guided by the perception that Usborne is a more 'youthful' community, with a younger population of farmers, accounting for higher rate of program adoption.

Census data however indicates that both Hobbs-McKenzie and Usborne have similar community age structures, with 35% of the population in both areas 55 years of age or older. The average farmer in Hobbs-McKenzie had 31 years of experience as compared to 27 years for Usborne. It was hypothesized that Hobbs-McKenzie's 'older' or 'aging' population would be less likely to invest in environmental voluntary programs. The validity of this hypothesis was nullified upon close examination of the data collected, thus requiring a closer examination of additional variables to account for poor program uptake in Hobbs-McKenzie.

After examining variables including farm size, gross farm sales and off-farm work, some interesting observations arose. Usborne has a greater percentage of medium farms in the 180-399 acre categories and also has a greater percentage of community farms than Hobbs-McKenzie, grossing between $120,000-$160-000 in sales per year. Consequently, Usborne has larger farms selling more products indicating that the agricultural community in this area is more economically stable. Also, a lower percentage (42%) of farmers in Usborne supplements their income with off-farm work. This compares to 44% in Hobbs-McKenzie. It was hypothesized that additional family income earned by farming families from off-farm employment would increase the feasibility for adoption of environmental stewardship programs, yet this was not the case in Hobbs-McKenzie. Although some valuable observations are noted by examining the relationship between program uptake and farm characteristics, a comprehensive understanding of the factors tied to program uptake requires further consideration.

As was shown to be the case in the nearby Grand River watershed (chapter 4) farmers within those rural communities with substantial non-farm populations such as Usborne are often more willing to participate in voluntary environmental programs and introduce more BMPs than communities comprised primarily of farmers like Hobbs-McKenzie. This may be a response to criticism from their non-farm rural neighbours who are concerned about agriculture's effects on the environment and who tend to support the greater regulation of agriculture inherent in the passage of the Nutrient Management Act (2002) and the Clean Water Act (2006).

Several studies by Bucknell (2002), Filson (1996), Howell and Laska (1992), Duff et al. (1991), and Arcury and Christianson (1990) have identified a link between level of education and environmental awareness; those who are more educated tend to have a greater concern for environmental conservation. Most Hobbs-McKenzie and Usborne respondents said that a high school education was their highest level of formal education while fewer had either some university, college or apprenticeship training. Therefore, it may be inferred that the level of formal education had little correlation or direct association with the decision to participate in the Healthy Futures program in the targeted, but our limited sample size makes it difficult to assess the importance of education on BMP adoption (see chapter 6).

Traore et al. (1996) corroborates the link between education and environmental awareness, contending that education to a key factor for program uptake and critical in stimulating landowner awareness of environmental degradation. There seemed to be a general lack of awareness and clarity by Hobbs-McKenzie respondents concerning what environmental stewardship programs and initiatives are available for adoption within their community and what organization(s) to contact regarding such programs.

Beyond discussion of proper manure spreading, many respondents in Hobbs-McKenzie and Usborne were unaware of what other environmental concerns may exist on their property or that their current practices may negatively affect environmental heath. Many landowners perceived themselves to be good 'stewards', having the notion that their current practices are 'best' practices.

As noted above, only 29% of Hobbs-McKenzie respondents identified the ABCA as a community resource for accessing information on stewardship programs, compared with 88% of Usborne respondents. Additional effort and measures are required on behalf on the ABCA to maintain a visible place within communities surrounding the Hobbs-McKenzie drainage area. Although no formal relationship could be determined between level of education and program uptake, the importance of including education and awareness building components into future programs must not be discounted. Increasing the public awareness of and inherent interest in environmental stewardship is critical to achieving a greater uptake of BMPs in any targeted community and also invaluable in the pursuit of increasing the community's capacity to manage local natural resources.A general tone of mistrust appears to exist between farmers and government agencies, specifically in the commentary provided by Hobbs-McKenzie participants. Changing the long established attitudes and perceptions of people provides an enormous challenge to the Ausable Bayfield Conservation Authority. The General Manager of the ABCA referred to this as 'historical hysteria', which in his estimation is based on past relations between landowners and the Conservation Authority. Although measures have been taken by the ABCA to rebuild the trust with landowners, the physical distance between the Hobbs-McKenzie area and the ABCA offices in Exeter presents an additional challenge.

Currently many farmers within Hobbs-McKenzie think that stewardship programs are the creation of urban technocrats who lack an understanding of the business of agriculture and the plight of farmers in general. Situated within a strongly rural, agricultural setting, the ABCA's General Manager has exercised an explicit agenda of staffing the Conservation Authority with individuals who have a strong agricultural background and in many cases live on farms themselves. While residents in Usborne may be more privy to this information based on their proximity to the ABCA, Hobbs-McKenzie residents have been left to assume that those delivering the stewardship programs possess a lack of sensitivity to farmers. These issues could be combated by issuing staff profiles in newsletters or local papers or through meeting CA staff in person at a meeting located nearby. Essentially, the farming community of Hobbs-McKenzie desires greater transparency and accountability of ABCA administration and programs. Local farm organizations can be a vehicle used to achieve this end.

Amicable, cooperative relationships with local farm organizations are an invaluable asset for any organization functioning within a rural community. Although General Manager Tom Prout notes that the ABCA maintains a 'professional relationship' with the Ontario Federation of Agriculture, the jurisdiction surrounding Hobbs-McKenzie has appointed a new representative with whom the ABCA has not had much interaction. The formation of a cooperative working relationship with the new OFA representative could be of great assistance in the delivery of environmental stewardship programs in future. Greater collaboration with local farm organizations can not only help to increase the validity of such programs in the eyes of the farm community but also act as an additional network for disseminating information.

Several comments were made by landowners within both target areas expressing dissatisfaction with the applications required to apply for stewardship program funds and the timeline under which financial reimbursement of project capital costs would take place. Further to this, many landowners felt that there is an inadequate amount of funding available for some programs with high capital costs and in cases where the community at large benefits from a BMP implemented, the cost to the farmer should be significantly less. This notion was corroborated by OFA representatives Paul Nairn and Dennis Bryson.

In times of economic uncertainty in agriculture, many farmers indicated that it is very difficult to find the money to pay for project capital costs up front, especially in some cases when reimbursement cheques were not issued for a year following project completion. This notion is concurrent with observations presented by Napier and Brown (1993) and Stonehouse and Bohl (1993) who contend that farmers are unlikely to engage in conservation practices if farm income could possibly decline as a result. This assertion is also consistent with the argument of Garling et al (2003) that determinants of intentions to perform pro-environmental behaviours include awareness of various environmental consequences, ascribed responsibility and personal norm. The ABCA's General Manager also expressed some frustration with inconsistencies in Environmental Voluntary Program's (EVP) administration, specifically emphasizing the difficulty that arises when different programs have different grant rates and guidelines.

Additional comments were provided by focus group participants regarding confidentiality and the application process association with EVPs. The mistrust of government agencies has left many farmers with a feeling of trepidation in volunteering information about questionable environmental practices on their farm and furthermore participating in programs where this type of disclosure in a mandatory component of the application process. However, the ABCA's General Manager indicated a system of coding has been applied to the filing system at the ABCA to ensure the anonymity of program participants and confidentiality of any information provided on grant applications.

Over the course of the data collection process, landowners were willing to candidly share their personal disincentives for program uptake. There is commonality in the lists of disincentives generated by both target groups indicating that many rural landowners share common concerns and face similar challenges. The most common disincentives for program uptake are:

- Process too lengthy and cumbersome (too much bureaucracy)
- Farmers feel threatened by 'negative' image of agriculture
- Mistrust of government
- Aging farming community
- Lack of money for project capital costs.

Similar problems were reported in McCallum's (2003) report discussing the barriers to farmer participation in agri-environmental programs. These results are:

Too many conditions might be attached:	28%
Too much time or paperwork:	26%
It might not be worth it economically:	13%
Don't like others involved in how I manage my land:	9%
Already have the best environmental stewardship:	7%

McCallum (2003) notes that there were often structural problems associated with the delivery of the design and delivery of the programs, insufficient financial support in addition to the attitudes that farmers had toward the programs, their mindscapes. Farmer resentment of the lack of funding and the requirement that they cost share (e.g. with Healthy Futures programs) was also found. McCallum pointed to the uncertainty created by the Nutrient Management Act which has been accompanied to some degree by a reduction in farmers' voluntary environmental behaviour.

The strong connection that farmers draw between economic farm viability and perceived quality of life creates an additional challenge in securing higher levels of program uptake. Some farmers expressed notable concern about low commodity prices and corresponding declining margins of operational profit. Selling farmers on the concept of environmental programs from the perspective of increased operational economic viability would perhaps evoke greater interest from the farm community in Hobbs-McKenzie. An increased availability of funding may ease the minds of farmers concerned with additional operational costs of implementing BMPs. Most importantly, the concept of paying farmers for the multifunctional agricultural purpose of producing environmental goods and services (EGS) as is happening in the European Union (EU), may well be seen as very attractive to farmers if it is ever implemented in Ontario. The additional prospect of cross-compliance, developed first in the EU and beginning to be implemented in Québec, also may eventually be seen in Ontario as governments work to find ways of protecting small farmers as well as the environment.

There are a number of incentives/motivations for increased program uptake that were suggested by landowners throughout the research process. Motivations can be categorized based upon the impact that it may produce for the landowner or farming operation specifically or where the action must come from for the incentive to be produced. These motivations have been synthesized into table 5.18 below. For example, motivators that involve new legislation or deal specifically with decision-making processes are categorized as 'political' and incentives that deal with financial stability of farms, cost effectiveness or improvements to operational efficiency are categorized as 'economic'. Motivational preferences differed between the two target areas; Hobbs-McKenzie landowners are more concerned with economic and administrative incentives and Usborne landowners are most interested in incentives that are political in nature.

Table 5.16: Motivations for Uptake of Voluntary Stewardship Programs

Hobbs-McKenzie	Usborne
Political Greater involvement of farm organizations (farmers) in process, equal decision-making between farmers/non-farmers about program guidelines and processes.	**Political** Knowledge of new legislation to be passed where action will be mandatory, Voluntary action with financial assistance is preferred.
Economic Greater level of financial compensation. Long-term programs with greater clarity of expectations and guidelines for the landowner.	**Political** Increased funding.
Administrative/Process Straight-forward application process.	**Social/Moral** Belief that it is the right thing to do
Administration/Process Assurance of confidentiality for information disclosed through program participation.	**Economic** Cost effectiveness or improvements in operational efficiency as a result of BMPs implemented.

CONCLUSION

Non-commercial environmental BMPs (those that do not cost much to implement) usually are implemented by small scale farmers whereas the large scale farmers are in a position to introduce commercial environmental BMPs as well. There are also farmers who farm almost full time and other farmers who are lifestyle or hobby farmers (usually with less than 130 acres and net income of less than $10,000) and this often limits the ability of the part-time farmers ability to implement BMPs. Both large operation and small operation farmers have different needs with respect to the kinds of BMPs that are usually implemented. Often times these 'lifestyle' farmers tend to have greater concern for the environment and often have more of an agrarian view of agriculture. Concern for water quality is often a solid reason for introducing BMPs for part-time farmers. People who are active participants in outdoor recreation, such as fishing and hiking, readily appreciate the value of introducing environmental BMPs. Thus it is always useful for conservation authorities to involve landowners in outdoor activities like stream walks, fly fishing workshops, etc. in their neighbourhoods as Grand River Conservation Authority agricultural extensionist Tracey Ryan has also observed (1999). This also has the benefit of reducing distrust of the conservation authorities among rural landowners.

The scale of operation and the type of farmer are important characteristics to take into account with regard to the nature of environmental BMPs to be implemented. Appealing to small scale farmers' interest in improving the environment makes sense for conservation authorities. For the larger scale farmer it is important to appeal to their interest in improving the profitability of their operation by pointing to the commercial advantages of implementing BMPs. With both groups it is important to stress that some of the benefits of introducing BMPs has a larger public benefit and it is therefore realistic to assume that some compensation from local conservation authorities for introducing the BMPs ought to be forthcoming (Ryan, 1999).

Whenever new financial and technical incentive programs are being considered it is essential to create learning platforms wherein the major farm organizations are encouraged to come to the table to help formulate the program. This strategy was successfully employed by Ryan when she initiated the Rural Water Quality Program (RWQP).

BMP participation in the two sub-watersheds was partly related to landowners' educational levels and their awareness of cost-share incentives to participate. Some nutrient management planning exists within these drainage areas but the extent of participation has been difficult to accurately assess. Adoption of BMPs is also partly related to the degree of farmer/non-farmer composition of each sub-watershed of the Ausable River.

The creation of agri-environmental policy and programs must be an iterative process, constantly evolving, requiring facilitators and participants to be flexible and able to adapt to contextual circumstances. Ontario has much to learn from progressive examples in the United States and European Union discussed in chapter three. A piecemeal 'one size fits all' solution to environmental issues has proven ineffective time and again because farmers have different mindscapes which in turn affect the EGS they will produce. The creation of future programs aimed at encouraging the adoption of BMPs will not be sufficient to sustainably manage our environmental resources and the agriculture sector. Ministerial policy-makers need to coordinate their efforts to produce integrative program frameworks that encourage good environmental practice on multiple levels, which takes into consideration the different landscapes as a result of landowners' perception about reality. Watershed-level environmental farm planning requires inter-agency, cross-jurisdictional collaboration and public participation. Past models of agri-environmental governance have largely embraced 'stovepipe' philosophies of management, unable to provide the flexibility, adaptability and level of collaboration required for effective implementation. As Aldo Leopold observed in 1949 (viii): "When the land does well for its owner, and the owner does well by his land--when both end up better by reason of the partnership—then we have conservation."

Chapter 6

CHANGING FARM PRACTICES IN SOUTHERN ONTARIO WATERSHEDS[1]

Glen Filson, Sridharan Sethuratnam, Barnidele Adekunle and Pamela Lamba

PURPOSE AND PERSPECTIVE

As we moved forward through the case studies discussed in chapters four and five we recognized the need to clarify some important relationships. We sought a larger sample of responses from farmers in several southern Ontario watersheds. We wanted to strengthen our findings' reliability and clarify the relationship between the numbers of best management practices (BMPs) farmers adopt and their gross farm sales, farm sizes, the types of commodities they produce, their watershed location, proximity to cities and conservation authorities and demographic characteristics. We also wanted to know why farmers adopt BMPs, what they think about the prospect of payment for environmental goods and services (EGS) and whether they think there should be cross-compliant regulations in return for receiving government payments

---

[1] This research is similar to two studies which were published as Filson, G. C., S. Sethuratnam, B. Adekunle and P. Lamba. 2009. Beneficial Management Practice Adoption in Five Southern Ontario Watersheds, *Journal of Sustainable Agriculture* 33(2): 229-252 and Lamba, P., G. C. Filson, and B. Adekunle. 2009. Factors affecting the Adoption of Best Management Practices in Southern Ontario, *The Environmentalist.* 29(1): 64-77. Unlike this chapter the *Journal of Sustainable Agriculture* article also incorporates data from a study by Lamba (2006) from Lake Simcoe region and Raison Region and it employs statistical techniques like correlation andanalysis of covariance which are not used here.

for agro-environmental activities. In the light of new regulatory measures like the Nutrient Management Act (NMA) and the Clean Water Act (CWA), questions pertaining to farmers' feelings about the relationship between voluntary and regulatory environmental management systems also needed to be addressed.

Ontario accounts for about one-fifth of the country's total agricultural production and nearly one-quarter of its agri-food exports. That has to be recognized and respected on both the national and international stages (National Farmers' Union, 2005). It is also true, however, that farming practices in southern Ontario have reduced surface and ground water quality in some watersheds. The imbalance of intensive conventional arable and livestock systems have increased the output of nitrogen, phosphorus and potassium and those plant nutrients often contaminate surface and ground waters. Intensive livestock units with poor waste handling facilities and insufficient land on which to apply these manures are important sources of phosphorus (Lampkin, 2002). This problem has become even more acute in southern Ontario where agricultural runoff is an important source of nutrients adding to problems of municipal and industrial effluents (Sly, 2000).

Environmental management programs have been a part of the Ontario agricultural production system for over 20 years. For instance, Ontario's Clean Water Program run by several southern Ontario Conservation Authorities such as the Ausable Bayfield Conservation Authority and the Grand River Conservation Authority's Rural Water Quality Program help to prevent environmental damage and improve environmental management.

BMPs are practical control measures that have been shown to effectively minimize water quality impacts (Ice, 2004). Thus when environmental incentive programs and regulations are developed it is important to understand what factors or conditions lead farmers' to adopt these BMPs. Farmers must be motivated to adopt and implement them appropriately because pro-environmental behaviour has a motivational component (Garling et al, 2003). Previous watershed studies have identified motivational factors affecting environmental adoption rates such as socio-economic factors (Arcury and Christianson, 1990; Howell and Laska, 1992; Serman, 1999), mistrust of government agencies (MacCallum, 2003), gross farm sales and farm income and age (Van Liere and Dunlap, 1981; Fransson and Garling, 1999; Lamba, 2006). As well, environmental best management practices (BMPs) have been adopted by many farm owners in various watersheds, but the rate of adoption is often not as high as required in order to reduce excess nutrients and other pollutants in some watersheds (Wells, 2004). Lamba (2006) concluded that many farmers were unsure of what the dollar benefits would be in the short run for implementing BMPs and as such often have waited to determine how performing these activities might benefit them financially. As well she found that many farmers were ill equipped financially to implement all necessary BMPs.

Jayasinge-Mudailige *et al.* (2005) noted the importance of proximity to urban areas as a major reason for why farmers adopt an environmental management system. Also, a number of studies have found that urban residents are usually more environmentally concerned than rural residents (Arcury and Christianson, 1990; Fransson and Garling, 1999). Consequently we also wanted to see if proximity to urban areas encouraged farmers to implement more BMPs.

As discussed in chapter three, Ontario farmers have been encouraged to identify any environmental risks of their operations by undertaking an Environmental Farm Plan (EFP). While the numbers of farmers who have implemented the EFP has risen substantially from the 20% Klupfel identified in 2000 to over half of all farmers (see Plummer *et al.*, 2007), many have only completed the EFP's first phase and some within the most fragile environmental areas have neither completed an EFP nor introduced appropriate BMPs to counteract environmental risks of their operations.

Justice O'Connor's *Report of the Walkerton Inquiry* concluded that stronger regulations were required to guarantee the quality of Ontario's drinking water. However, Lamba (2006) found that most farmers already feel that there is too much government regulation and many resent recent environmental legislation requiring them to have a Nutrient Management Plan (NMP) and/or other environmental programming. McCallum (2003), Agnew (2004) and Wells (2004) have identified the perceived barriers agri-environmental programs must address in order to be implemented by Ontario farmers but greater clarity is still required about who is most in need of doing BMPs most and what the relationship of BMP adoption is to voluntary/regulatory programming, gross farm sales, farm size and location as well as demographic variables affecting farmers' mindscapes.

Beaulieu (2005) used logistic regression to model the likelihood of implementing BMPs based on farmer and demographic characteristics. The age group among farmers that he found to be implementing the most BMPs was 45-54 years. Full time operators were also discovered to have adopted more BMPs than part-timers. Larger, more specialized farms had implemented the most BMPs, but income was not found to be related to the BMP adoption rate. At times farmers' education has been identified as a relevant factor affecting adoption rates (see chapter 4) but in other watersheds surveyed at different times, it seems not to be important (Lamba *et al.*, 2009). The decision making process which a farm owner undertakes is complex as they have to put these practices in place at different levels of their operation at different times of the year. Thus there are many complex interactions between farmers' social and technical relations which impact on their decisions to adopt BMPs (Leeuwis, 2004).

Farmers' social relations and perceived social pressures which affect them plus their technical and social practices also generate feedback from their agro-ecological and social world which affect them, in turn impacting their perceived environmental effectiveness and efficacy, moving them forward into somewhat revised technical and social practices as the cycle continues. The combined effect of voluntary environmental program participation and appropriate regulations ought to increase their degree of environmental friendliness over time.

RESEARCH METHODS

Prior to the surveys that generated the empirical data for this chapter we considered our most important hypotheses based on the literature and our previous findings. For instance, we thought that the higher the farms' gross farm sales, the higher would be their Adoption Rate Index (ARI)—based on the numbers of BMPs they have adopted. We also expected that the larger the farm, the higher would be the ARI. Secondary hypotheses were developed relating the adoption of BMPs (the ARI) to their location within watersheds and as well as such demographic variables as farmers' educational level, age and gender.

These were then tested through the development of an Adoption Rate Index (ARI) based on the numbers of BMPs the farmers said they had adopted. Social profiles of the farmers and their types of agriculture were then constructed from Statistics Canada's agricultural census data for each sub-watershed. Descriptive and inferential statistics were then used to analyze the data. This included the development of two logistic regression models designed to predict which factors affected adoption rates the most.

The primary data for the study were collected through the use of farmer mail-out-questionnaires. The social profiles were used to examine the agricultural activities in each sub-watershed and the highlights of these social profiles are mentioned below. The sample was obtained using stratified random sampling selected in proportion to the number of farmers in the best and worst sub-watersheds in Lake Simcoe, Grand River and Ausable Bay similarly to previous surveys in Maitland Valley and Raison Region (Lamba, 2008).

Once the political boundaries were identified within each sub-watershed, random samples were selected so that sampling error for the populations within the sub-watersheds be less than 10%. Questionnaires were then mailed to farmers in the various sub-watersheds.[2] Table 6.1 provides the approximate sampling errors of returned questionnaires in each watershed.

Table 6.1: Sample Sizes and Sampling Error

Watershed Sub-watershed ..	Sample size	Questionnaires returned	Sampling error
(For both Sub-watersheds)			
Ausable Usborne	587-3=584*	83	+/- 10%
Bayfield Hobbs-McKenzie			
Grand River Upper Nith	1009-45=964*	174	+/- 6.5%
Eramosa-Speed			
Lake Simcoe Beaver	1012-38= 974*	56	+/- >10%
Black			
Unknown Watersheds 4			
Total	2252	317	+/- 5-6%

DATA ANALYSIS

Descriptive statistics such as means and frequencies were used to describe some of the farm and respondent characteristics in the study. Logistic Regression was employed to determine the variables that predict the adoption rate index (ARI), which is a measure of adoption of BMPs by farmers based on their total number of adopted BMPs. Logistic regression allows one to

---

[2] A substantial number of questionnaires were returned because of incorrect addresses, owners no longer farming or landowners' deaths. Many of these addresses had not been removed from the municipal assessment rolls used as the sampling frame, which contributed to the sampling error.

predict a discrete outcome, such as high or low adopter, from a set of variables that may be continuous, discrete, dichotomous, or a mix of any of these. Generally, the dependent or response variable is dichotomous, such as presence/absence or success/failure. The independent or predictor variables in logistic regression can take any form.[3] T tests and ANOVA were also used to test for significant differences in means. Correlations were calculated to test for significant relationships among the variables.

Social Profiles of Farmers in the Watersheds Being Studied

Lake Simcoe Watershed (middle of Figure 1.1)

The reason the Beaver River (Durham Region) is considered the most degraded sub-watershed by Lake Simcoe Conservation Authority officials could be because of the amount and types of agriculture in the area including such livestock as beef (the highest relative percentage of the provincial total), followed by sheep and goats, then dairy and chickens/turkeys followed by a relatively small percentage of pigs. There is also significant crop production as well but each type, the biggest of which is mixed grains, followed by barley and oats, are of a smaller percentage than elsewhere in the province. The major field crops tend to be mixed grains and barley. About 87% (1451) of Beaver River farmers reported gross farm sales (GFS) of less than $249,999 whereas the remaining 13% (218) had GFS above $250,000 per year (Statistics Canada, 2001).

The Black River is located in the York Region and is considered by the local Conservation Authority (CA) to be the least degraded sub-watershed in the Lake Simcoe area. This may be due to the reduced amount of agriculture in the area compared with the Beaver Creek sub-watershed. There are only 1,020 farmers and only 95 of them occupy more than 162 hectares each. York's percentage of livestock is relatively quite small. The largest relative provincial numbers for York are for sheep and goats followed by chickens, beef, dairy and pigs. Potatoes, then corn for silage and grain followed by mixed grains and winter wheat are the largest crops but they're relatively insignificant provincially. Only 15.7% have gross farm sales (GFS) above $250,000 in

---

[3] That is, logistic regression makes no assumptions about the distribution of the independent variables. They do not have to be normally distributed, linearly related or of equal variance within each group. The relationship between the predictor and response variables is not a linear function in logistic regression.

York and there are substantially fewer farms in this watershed than in Durham Region (Statistics Canada, 2001).

Ausable Bayfield Watershed (see Figures 1.1 and 5.2)

As revealed in Chapter 5, Hobbs-McKenzie Drainage Area is perceived by members of the Ausable Bayfield Conservation Authority to be among the most degraded, whereas Usborne sub-watershed about the best. The boundaries of the Hobbs-McKenzie Drainage Area are delineated on a watershed basis and therefore do not follow the geographical boundaries outlined by Census Canada. The Hobbs-McKenzie Drainage Area follows the natural course of the Ausable River, which traverses more than one county and township making it difficult to definitively use Statistics Canada data to fit explicitly with this drainage area. Although some property owners within the Hobbs McKenzie (H-M) area are technically residents of the Municipality of Lambton Shores, Warwick Township with more H-M property owners was thought to be a more appropriate census area from which to build the community profile.

Usborne Township has 1,535 residents and therefore about one-third of the population reside in Warwick Township (4,025 residents). There are 445 people engaged in agriculture in Usborne. Overall, Usborne has a greater proportion of the total population in agriculture at 29% compared to 10% in Warwick. Although both areas are considered rural, almost 70% of the Usborne people are non-farmers while almost 90% of the residents in Hobbs-McKenzie are non-farmers.

Both communities have a comparatively similar cross-section of farms by size category, mostly small to medium sized operations. Warwick Township has a greater number of recreational or hobby farms.Though Usborne has slightly more farmers than Hobbs-McKenzie the approximate percentage of farm size is similar with 36.7% of Usborne farms being larger than 400 acres compared with 33.9% of Warwick's farms..

Grand River Watershed (see Figure 4.1)

Next to the Canagagigue Creek, the Upper Nith watershed is considered by local CA officials to be the most degraded sub-watershed in the Grand River watershed[4]. The census data was extracted by consolidating the census sub-divisions of Wellesley and Wilmont townships. Upper Nith sub-watershed

---

[4] Having recently conducted a survey in Canagagigue Creek (CC), we felt that it was wise to change the sub-watershed so as to provide a break for the CC farmers.

had a total population of 24,231 of which only 1,025 (4%) were agricultural operators. The total number of farms reported in this subwatershed was 772 with an average farm size of 160 acres. Compared with the average farm size in the Grand River watershed that there are a high proportion of small farms in the watershed. Still, there is a significant percent (61%) of livestock farms and there and farms (26.9%) with gross farm sales of over $250,000 (26.9%) (Statistics Canada, 2001).

The census data for the sub-watershed with much better water quality was collected from the Guelph-Eramosa Township (previously Eramosa Township) and Erin Township. The total population in this watershed was 128,386 out of which only 765 were classified as agricultural operators. The total numbers of farms were 765 with 174.5 acres being the average acreage (Statistics Canada, 2001).

In Eramosa/Speed and Erin and unlike Upper Nith both livestock and field crops are fairly evenly distributed. The highest percentage of farms is in the miscellaneous specialty farms category (26%) with 23% cattle, 15% grains and oil seed and only 11% dairy among other types. In Eramosa-Speed only 14.8% of the farms have gross farm sales in excess of $250,000 compared with 26.9% in Upper Nith (Statistics Canada, 2001).

Respondents' Farm and Demographic Characteristics

The farmers sampled were predominantly male (81.2%). Pariticipants averaged 54 years with the minimum age being 21 years and the maximum age being 86 years old.

The average farm size within our sample was 203 acres.[5] Almost 55% of our respondents stated that off-farm income contributed significantly to their total income. Eighty-three percent of our respondents had more than 20 years of farming experience out of which 29% had farmed for more than 40 years. On average respondents said that their perceived quality of life had declined from 74.2% indicating good or very good in 2001 to 60.2 percent in 2006.

---

[5] More specific tabular details about the livestock types and numbers, field crops, total gross farm receipts, numbers and percentage of farms, farm sizes, respondents' roles on the farms for Durham and York Regions, Eramosa/Speed and Erin Township, Warwick and Usborne can be found in Filson and Agnew, 2004 and Sethuratnam, 2007.

Farmers' Perceptions of Environmental Programs

When asked about the environmental program involvement, 143 out of 224 (63.8%) said that they had participated to some degree in the Environmental Farm Plan (EFP). The next highest level of participation was in Land Stewardship I and II (55%) about one-quarter.

Of the two-thirds who were aware of the idea of farmers being paid to produce environmental goods and services (EGS), 90.5% said that the EGS concept was a good idea. Most (51.4%) considered EGS to be a method of receiving payment for producing environmental goods and services and the second highest percentage (29.2%) thought that EGS was farmers' contribution to the public good. One woman commented that

If there is a place for Ecological Goods and Services wherein the public can partake, without adding to the tax bill, without benefiting those who take without giving in society and whose mandate is clearly stated, understood by all, then we would probably support it. We do believe in having our land remain productive; with little or no erosion, in having clean water (we use a cistern and roof rainwater for almost everything but drinking) and will continue to do our best to follow anti-pollution rules and regulations.

When asked what their reaction was to government regulation in general, 135 out of 286 (47.2%) indicated that there was already too much regulation of farmers while 102 (35.7%) thought that the level of regulation was adequate. Only 44 (15.3%) felt that the level of regulation was adequate. Regarding the recently passed Clean Water Act (2006) which mandates source water protection 221 out of 302 (73.2%) were unfamiliar with this regulation. Farmers were also asked to explain how they felt about the revised Nutrient Management regulations. Almost half (44%) felt that the regulation revisions were necessary but one-third thought that, even after revision, the Nutrient Management Act (NMA) is still too restrictive.

Government support of some of the costs of environmental programs is a major issue among farmers. A significant percentage of farmers both small/medium and large, indicated that they would like to receive more than 50 percent funding from the Government. About 56% of the small to medium farmers would like to receive 50% or more support from the government and whereas only 39% of the larger farmers felt they needed more than 50% of the cost funded by the Government supported.

We asked farmers if they felt that it was reasonable for the government to expect cross-compliance whereby farmers would not receive property tax reimbursements, or financial support unless they complied with environmental regulations. Again we found that 36% of the respondents felt that there is a need for more financial support and incentives for cross in order for-compliance to work. About one-quarter of them (26%) felt that there is too much government involvement already and therefore they would not support cross-compliance requirements, such as exist, for example, in Europe.

Farmers were asked how regulations could be targeted to achieve the necessary safeguards for the environment allowing maximum flexibility and imposing the least burden on farmers. Once again the largest number (39%) felt that there needs to be more financial support/financial incentives given to farmers. Twenty-nine percent thought that education and training should be provided to farmers.

Finally, we asked farmers how these environmental farm practices including Environmental Farm Planning, Nutrient Management, the Source Water Protection, Clean Water Act, etc. affect their quality of life. The largest percentage (41%) stated their quality of life has improved for the better as the result of increasing environmental practices. On the other hand, 34% felt that there is more financial burden and stress in their life since these regulations and environmental programs have been implemented.

Adoption Rate (ARI) Logistic Regression Model

This study examined the factors that might affect the adoption rate of BMPs by farmers in the study area. The model below was specified based on our earlier case studies, participant observation and in-depth interviews.

Two regression equations were created to explain the relationship between the independent and dependent variables. The first of these equations/models was:

Adoption Rate = f(education, off farm income, gender, age, proximity to city, proximity to conservation authority, watersheds, total land farmed)

With this model, the results indicated that only the size of land farmed affects the adoption rate, and all the other factors do not have a significant effect on the adoption rate of the farmers in the study area. If the size of farm increases by one unit the odds of high adoption will increase by 1.002 times (Table 6.2).

The Nagelkerke R^2 of 0.188 indicates that 18% of the variation in the dependent variable is explained by the explanatory variables in this model. This is low because a low R^2 is a normal occurrence in cross sectional study of this type but it does not have a negative effect on the result.

Table 6.2: Effects of Different Explanatory
Variables on Adoption Rate

	B	S.E.	Wald	df	Sig.	Exp (B)
Watersheds	-.316	.488	.418	1	.518	.729
Age	.018	.019	.925	1	.336	1.018
Gender	.576	.815	.501	1	.479	1.780
Off-farm income	.045	.519	.008	1	.930	1.047
Education	-.966	.716	1.820	1	.177	.381
PRCA	-1.613	.886	3.314	1	.069	.199
PRCT0	.864	.583	2.197	1	.138	2.373
Total land farmed	.002	.001	7.858	1	**.005**	**1.002**
Constant	-2.438	1.49	2.659	1	.103	.087

This study used annual gross farm sales as a measure of performance. The results as indicated in Table 6.3illustrate that age, off farm income, total size of land farmed and adoption rate index predict gross farm sales. Total farm size and adoption index have the highest magnitude. If the land increases by one unit the odds of high sales will increase by a factor of 1.011. In the case of adoption rate, an increase in the adoption rate by a unit will increase the odds of high sales by 180.80 times. It can thus be deduced from this result that farmers with good management practices (higher adoption rates) and large farm sizes are likely to have better sales. Thus we should try to encourage farmers to adopt BMPs since it may be connected to higher farm sales though the nature of the causality between these variables is unclear.

Gross Farm Sales (GFS) Model

Gross Farm Sales = f(education, off farm income, gender, age, best and worst sub-watersheds, total land farmed)

The Nagelkerke R Square of this model is very good. The R^2 has a value of 0.708 which suggests that 70.8 percent of the variation in the gross sales of the sampled farmers is explained by this model.

Table 6.3: Factors that Affect Gross Sales

	B	S.E.	Wald	ddf	Sig.	Exp (B)
Watersheds	-.277	.429	.417	1	.519	.758
Age	-.072	.018	15.262	1	**.000**	**.930**
Gender	-.052	.659	.006	1	.937	.949
Off-farm income	-2.358	.459	26.424	1	**.000**	**095**
Education	-.187	.572	.107	1	.744	.829
Total land farmed	.01	.00	23.840	1	**.000**	**1.011**
Adoption rate index	5.197	1.384	14.108	1	**.000**	**180.801**
Constant	1.501	1.17	1.623	1	.203	4.487

Results of T Tests

The T test was used to determine whether there was significant difference in the ARI of different groups within different variables that are characteristic to the farmers in the study area.

a) Farm Size

The mean value of ARI for farms less than 299 acres was 0.24 and for farms more 300 acres or more was 0.42. The T test at the 5% level of significance shows that there was a highly significant difference in the means between large farms and small farms. Thus, larger farms are more likely to adopt management practices.

b) Gross Farm Sales

The mean value of ARI for farm sales less than $49,999 was 0.19 and for farms with gross sales $50,000 or more was 0.38. The T test at 5% level of significance shows that there was a highly significant difference in the means between farms with higher gross sales and lower gross sales. It can thus be deduced that the higher the gross farm sales the higher the adoption rate of farmers.

c)   Environmental Farm Plan (EFP)

The mean value of ARI for farmers who had implemented the EFP was 0.36 and for those who had not implemented it was 0.21. The T test at 5% level of significance shows that there was a highly significant difference in the means, which shows that those who had implemented an EFP had higher adoption rates than those who had not.

d)   Nutrient Management Planning (NMP)

The mean value of ARI for farmers who have a NMP was 0.45 as against 0.26 for those who do not. T test at 5% level of significance shows that there was a highly significant difference in the means, which indicates that farms which adopted NMPs had a higher adoption rate of BMPs.

e)   Watersheds (least and worst degraded)

The mean value of ARI of farmers in the most degraded watersheds (Upper Nith[6], Black River and Hobbs-McKenzie) was 0.26 and of those in the least degraded (Beaver River, Eramosa-Speed and Usborne) was also 0.26. The T test at 5% level of significance illustrates that there was no significant difference in the means between the worst and least degraded watersheds.Therefore, this study did not find any direct link between the quality of water in particular sub-watersheds and farmers' implementation of BMPs.

f)   Proximity to a City

The mean value of the ARI for proximity to cities with more than 50,000 residents less than 49 km away was 0.26 and that of proximity to cities more than 50 km was 0.27. T test at 5% level of significance shows that there was no significant difference in the means between farmers close to or far away from cities.

---

[6] The water quality in Canagagigue Creek (CC) is actually worse than that of the Upper Nith but the researcher's interviews with CC farmers in 2005 ruled out using this subwatershed again (see chapter 5).

g) Proximity to Conservation Authority

The mean value of ARI for farmers close to conservation authorities (less than 39km) was 0.27 and further away (more than 40 km) was 0.43. T test at 5% level of significance shows that there was a significant difference in the means; this suggests that farms far away from conservation authorities had a higher adoption rate. This should be interpreted with care, since in the sample only eight were in the more than 40 km group as compared with 193 in the other group (less than 39km). Apart from the problem of unequal group all the eight farmers might be high adopters.

h) Education

The mean value of ARI for education levels up to college/apprenticeship after high school was 0.28 and for university level education was 0.23. The T test at 5% level of significance shows that there was no significant difference in the means. Thus farmers' level of education had no affect on adoption rates in the areas sampled here.

i) Age

The mean value of ARI of the respondents who were young (0 to 40 years) was 0.26 and that of those who were older (41 years and above) was 0.27. T test at 5% level of significance shows that there was no significant difference in the means between the younger and older farmers, indicating that age did not affect farmers' adoption rate. There is a statistically significant inverse correlation between age and ARI but it is only -0.167. Again this differs from the earlier finding with a different sample that showed younger people being more inclined to introduce BMPs.

j) Gender

The mean value of ARI for males was 0.25 and for females was 0.18. The T test at 5% level of significance shows that there was a highly significant difference in the means, which indicates that men are higher adopters than women. A note of caution should be introduced as the result of the fact that the sample size between men (251) and women (58) was highly disproportionate.

k) Off-Farm Income

The mean value of ARI for farmers who stated that off-farm income did not contribute to their total farm income was 0.31 and those that stated that it did was 0.26. The T test at the 5% level of significance shows that there was a significant difference in the means. Farmers without off-farm income are better adopters.

A frequency test was run to determine the main reason for why participants adopt BMPs. Farmers ranked environmental responsibility the number one reason at 58 percent followed by financial incentives at 18%, economic benefits at 15% and concern for environmental regulation at 14%. This indicates that farmers care for and have a sense of responsibility towards protecting the environment.

A bivariate correlation test was done to determine the relationship between the farmers' perceived quality of life and the adoption rate. As seen in Table 6.7, there is a positive correlation between the adoption rate index and quality of life (2001), while the relationship between adoption rate and quality of life (2006) is not significant. Thus farmers' views of their quality of life have been changing.

Analysis of Variance (ANOVA) was done to determine whether there was asignificant difference in ARIs average based on type of farming practiced (see Table 6.8).

Table 6.4: Relationship between ARI and Quality of life

	Pearson Correlation (2-tailed significance)	Adoption rate index
Adoption rate index	Significance	1
	N	317
Quality of life (2006)	Correlation	-.008
	Significance	.897
	N	299
Quality of life (2001)	Correlation	.155(**)
	Significance	.008
	N	295

**   Correlation is significant at the 0.01 level (2-tailed).

Table 6.5 Adoption Rate Index and Type of Farm Commodity

Farm Commodity	N	Mean
Dairy	39	.3910
Poultry	13	.3531
Beef	31	.3400
Swine	25	.3300
Crops	121	.2736
Small Ruminants	7	.1957
Horses	3	.1700
Total	239	.3080

The means were significantly different at the 5% level of significance (value 0.008). Thus, table 6.5 indicates that livestock frms were high adopters, as they should be to avoid excess nutrients in these watersheds. Dairy operations, for instance, are generally full time operations which are reasonably secure financially and so it is not surprising that they are usually the highest adopters. Also, the Nutrient Management Act mandates that large animal operations (over 300 Animal Units), must undertake NMPs and incorporate BMPs, which of course is much larger than the average dairy farm which had only about 65 cows as recently as 2003 (Pfeiffer and Filson, 2004).

The correction matrix of the variables is presented in Table 6.6.

LIMITATIONS OF THIS ANALYSIS

For convenience, we've treated each of the 21 best management practices making up our Adoption Rate Index (ARI) as if they are equally important for all farmers. This is problematic because of the difference between the BMPs adopted primarily by cropping based farming systems, for example, and the BMPs adopted mainly livestock farming systems. A more exact assessment of the ARI would have distinguished between the farming system and the BMPs adopted.

Also, the ARI does not take account of the differences in the value of specific BMPs but is instead based entirely on the total number of BMPs implemented. Future research will have to find a way of distinguishing among the quality as well as the number of BMPs.

Table 6.6: Correlations among the Variables

Variables	Pearson Correlations/ Significance/N	Watersheds	Age	Gender	Off-farm Income	Education	Total Land Farmed	ARI	Gross Farm Sales
Watersheds	Pearson Correlation	1	.105	-.054	.170 (**)	.066	-.096	-.001	-.115 (*)
	Sig. (2-tailed)		.077	.357	.005	.263	.132	.991	.050
	N	292	283	289	275	285	247	292	292
Age	Pearson Correlation	.105	1	-.032	.066	-.006	-.177 (**)	-.167 (**)	-.325 (**)
	Sig. (2-tailed)	.077		.581	.270	.915	.005	.004	.000
	N	283	297	297	282	293	255	297	297
Gender	Pearson Correlation	-.054	-.032	1	-.066	-.153 (**)	.129 (*)	.193 (**)	.216 (**)
	Sig. (2-tailed)	.357	.581		.259	.008	.037	.001	.000
	N	289	297	309	292	301	262	309	309
Off-farm Income	Pearson Correlation	.170 (**)	.066	-.066	1	.273 (**)	-.278 (**)	-.129 (*)	-.499 (**)
	Sig. (2-tailed)	.005	.270	.259		.000	.000	.027	.000
	N	275	282	292	295	287	256	295	295

Table 6.6. Continued

Variables	Pearson Correlations/ Significance/N	Watersheds	Age	Gender	Off-farm Income	Education	Total Land Farmed	ARI	Gross Farm Sales
Education	Pearson Correlation	.066	-.006	-.153 (**)	.273 (**)	1	-.154 (*)	-.107	-.182 (**)
	Sig. (2-tailed)	.263	.915	.008	.000		.014	.064	.001
	N	285	293	301	287	303	258	303	303
Total land farmed	Pearson Correlation	-.096	-.177 (**)	.129(*)	-.278(**)	.154 (*)	1	.347 (**)	.499 (**)
	Sig.(2-tailed)	.132	.005	.037	.000	.014		.000	.000
	N	247	255	262	256	258	263	263	263
ARI	Pearson Correlation	-.001	-.167 (**)	.193 (**)	-.129 (*)	-.107	.347 (**)	1	.457 (**)
	Sig. (tailed)	.991	.004	.001	.027	.064	.000		.000
	N	292	297	309	295	303	263	317	317
Gross Farm Sales	Pearson Correlation	-.115 (*)	-.325 (**)	.216 (**)	-.499 (**)	-.182 (**)	.499 (**)	.457 (**)	1
	Sig.(2-tailed)	.050	.000	.000	.000	.001	.000	.000	
	N	292	297	309	295	303	263	317	317

** Correlation is significant at the 0.01 level (2-tailed).
* Correlation is significant at the 0.05 level (2-tailed).

CONCLUSIONS

This study shows the importance of gross farm revenue, farm size in determining farmers' propensity to introduce BMPs. The implementation of EFP and an NMP, full time farming and operating a relatively large livestock operation are also correlated with a high adoption rate of BMPs. On the other hand, the environmental quality of the watershed, proximity to cities, age and education did not affect the adoption rate significantly. Proximity to a Conservation Authority (CA) had a significant relationship but close inspection of the mean responses shows that the small sample of those at a great distance from a CA makes this result unlikely.

Financial incentives are can encourage adoption of BMPs and EFPs especially when adoption occurs at a net cost to farmers. Clearly farmers feel a significant responsibility towards the environment though most strongly oppose environmental management by regulation. Farmers in fragile environmental areas who do not voluntarily adopt an EFP and implement BMPs should nevertheless comply with existing environmental regulations. Some regulation and monitoring is required but too much regulation may also have detrimental effects on adoption rates.

Farmers felt that their perceived quality of life had declined since 2001. The study showed that many farmers' access to information about the NMA and Clean Water Act has been inadequate. This may be due to OMAFRA's reduced number of extensionists as it tries to provide information to farmers primarily through a call centre and its website. Extensionists working with the CAs have done a terrific job in all three watersheds but the CAs are still woefully under-funded as well.

In sum, this study showed that farmers with good management practices (higher adoption rates) and large farm sizes are clearly likely to have better sales. It is obvious that the watersheds in this study have a large proportion of small farmers and this explains why the overall adoption rate is low at 0.27. In order that more small farmers participate, programs must be designed so that they target small farmers to encourage them to adopt environmental BMPs.

Chapter 7

TOWARDS A MORE ENVIRONMENTALLY SECURE AGRICULTURE

Glen Filson and Bamidele Adekunle

INTRODUCTION

This chapter considers the case study findings presented earlier in relation to the larger international context prior to developing recommendations to increase agro-environmental sustainability. Utilizing a perspective derived from the most likely determinants of agro-ecosystem environmental security, we have studied the centrality of farmers' attitudes and motivations regarding how their management practices interact with and affect their environment. We measured landowners' environmental attitudes and motivations in relation to various environmental programs which are generally designed to encourage the adoption of environmentally beneficial management practices (BMPs).

Clearly the effects of the latest globalized international food regime, developments in technology and the intensification of agriculture in southern Ontario have raised productivity while jeopardizing environmental and social sustainability. Simultaneously a transition has been occurring from family farming which functioned without hired labour, except occasionally, to capitalist farming which employs farm labour. Family farmers are sliding into the worker farmer category or disappearing and even medium sized capitalist family producers are increasingly employing hired labour while many others have left farming altogether.

The continued destruction of small operation farming has also threatened social sustainability and tipped the balance of municipal control toward non-farmers in many southern, rural Ontario communities. The expansion of urban sprawl in the countryside is being abated somewhat by recent Greenbelt zoning legislation but this too angers many farmers within the areas protected from development via this Act whose property has diminished in value as a consequence.

Ontario's environmental sustainability is at risk because of greater soil compaction[1], increased dependence upon hormones and pharmaceuticals in livestock production, excess nutrients in some watersheds including *E. coli bacteria* O157:H7 which has acid resistance, in part because ruminants have been fed a high corn diet (Kenner, 2009). This form of *E. coli* bacteria from a beef herd which has not been destroyed by sufficient chlorine was what had caused 7 deaths and injured over 2100 people in the Walkerton tragedy of 2000. Our industrialized agriculture is also so hooked on fossil fuels such that for the first time in history, it now consumes more energy than it produces (Kirschenmann, 2008), in turn contributing to climate change. There is also well contamination, loss of wetlands and reduced biodiversity (Filson, 2004d). Compounding the negative image of the agri-food industry recently was the contamination of cold cuts of meat in Maple Leaf Foods' Toronto processing plant which created a *listeriosis* outbreak which killed 20 people in 2008[2].

While this book's watershed approach to case studies does not deal with either the mitigation or adaptation to climate change directly,[3] some of the BMPs which are being implemented by many Ontario farmers, discussed in chapters 4-6, are slowing the production of greenhouse gases (mitigation) while others contribute toward climate change adaptation. Progress toward greater EFP participation and also the adoption of BMPs does enhance the prospects for sustainable development but the argument for branding Ontario grown food as environmentally friendly due to the progress that has been made to date is clearly premature. But before pointing to our findings' significance

---

[1] A 1990 MAPAQ study in Québec showed that soil compaction was mainly due to intensive cropping (particularly of corn and potatoes) causing compaction, erosion, reduced organic matter and soil aggregate damage. The situation in Ontario is similar.

[2] The fact that another 20 people died from *listeriosis* caused by eating food (in this case cold cuts of meat) containing bacterium *Listeria monocytogenes* also has been blamed on the Federal Conservatives by Dr. Carolyn Bennett, a Liberal health critic. Stéphane Dion, the former Liberal Leader in September, said "Starting March 1, a change has been made that put our inspection situation where inspectors are more inspecting paper than meat" (Whittington, 2008).

[3] Dealing with animal welfare was also deemed to be beyond the scope of this book.

with respect to future agro-environmental policy let us outline the main elements of the argument so far.

TOWARDS AN AGRARIAN POLITICAL ECONOMY OF ONTARIO AGRICULTURE

The political economic approach used here enables the connection of the environmental impacts with the economic structure of agriculture. It also helps to elucidate political forces that shape environmental policies, facilitating the ability to track their environmental foundations (Hessing *et al.,* 2005). The political economic approach was chosen because of its ability to show how environmental and social sustainability is limited by the will and ability of governments in the face of producer pressures from large agri-food companies and farmers who do not speak with one voice due to the conflicts of interest among different farming classes.

Along these agrarian political economic lines, for example, Montpetit argued in 2002 that farmers have controlled the environmental agenda of Ontario generating the 'Normal Farm Practices' Law protecting farmers from nuisance complaints and self regulation of their environmental practices until the Walkerton Tragedy. This 'clientelism' collapsed once the provincial Conservatives were thrown out of office and the Liberals passed several environmental regulations including the Nutrient Management Act (2004), the Clean Water Act, 2006 and the revised Endangered Species Act (2009).

The agrarian political economic perspective works best in analyzing changing food regimes and the increasingly globalized nature of the capital invested in agri-food systems, when combined with dialectical systems thinking. This concentrates much of its attention on production relations both within farming and between different levels of agri-food business and farmers though the threat to the globalized food system described in the paragraph above, also points to the importance of food distribution in addition to food production. Nonetheless, the transformative moments within the realms of production are crucially important to the understanding of the degree of sustainability of agriculture. As an important input into the production process, circulating capital including oil, can not only trigger a recession when it spikes, it can also dramatically increase food prices leading to food riots in 2007 and 2008 around the world.

It is useful to try to understand these systems (food regimes, farming systems, production/distribution systems, class structures, etc.) in a dialectical fashion because because understanding internal and external contradictions within, between and among systems of production, governance, food processing and production helps us understand the changes that are occurring.

For instance, the past few years of recession were triggered by a financial meltdown linked to the sub-prime mortgage boondoggle in the U. S. but ultimately in turn to the overproduction of housing, automobiles and other consumer durables in the industrialized capitalist countries relative to effective demand. Prior to the collapse oil had reached $147/barrel U.S. at a time when food prices were also at a peak. Given the extent to which industrialized agriculture and food processing utilizes fossil fuels, the price of food has tracked oil prices fairly closely. While the price of corn, wheat and other food staples has come down somewhat since, the steady growth of demand from the rising middle classes of China and India where growth continued unabated has kept food prices firmer than one might otherwise have expected. Rubin (2009) even argues that the primary cause of the 2007-2009 recession was triple digit oil prices and not the excesses of Wall Street. He believes that because we are at or nearly at peak oil, the point where more oil is consumed per year than is discovered, we will continue to see even higher oil prices. Among other things, this will add strength to the local food movement because of the excessive cost of shipping food around the world as we have been doing in the latest food regime.

With this theoretical background, the book then turned to case studies of farmers' environmental management systems in chapters 4-6.

LESSONS LEARNED FROM WATERSHED STUDIES

Farmers' environmental attitudes, behavior and perceived quality of life were studied in the Grand River (chapter 4) based on an analysis of focus group meetings and questionnaires completed by farmers and non-farmers in the most degraded sub-watershed and the least degraded sub-watershed. Because there are fewer water quality problems in the Eramosa/Speed (E/S) region than in the Canagagigue Creek (CC), we wanted to understand why. We thought this might be due to different land use practices, degrees of agricultural intensity, the attitudes of rural people toward BMPs and the impact those practices might have on rural people's perceived quality of life. We also wanted to compare residents' views about the voluntary adoption of BMPs versus the post-Walkerton tragedy requirement that large operation farmers implement nutrient management planning. As elsewhere in southern Ontario, there was a widespread rural mis-perception that a few large farm operations are a greater environmental threat than many small farms.

Most of the Grand's farmers strongly prefer voluntary programming and resent what they perceive to be growing government intervention in agriculture. They believe that the intervention interferes with their independence, increases their workload and costs them scarce money and labour. This is because government mandated programs usually do not include all of the costs even when such activities are required under the NMA or CWA. Extended negotiations with groups like the OFA, the CFFO and OFEC have nevertheless led to changes in these regulations which have made them more palatable to farmers.

The relatively low level of awareness of and participation within government/CA environmental programs among many small farm operators in CC is connected to the proliferation of livestock operations in the area and, on average, a limited formal education. Nevertheless, a prominent member of the OSCIA has observed that this is changing as many CC farmers are now coming, often in groups, to OSCIA EFP meetings.

The EFP is required for farmers to qualify for the GRCA's RWQP which provides up to 100 percent of the costs of some BMPs where these would otherwise come at a net cost to farmers. Also providing greater incentives to farmers who have completed the EFP is the recently developed Clean Water Program (CWP) which is run not only by the GRCA but by Ausable Bayfield, Catfish Creek, Kettle Creek, Long Point Region, Lower Thames Valley, Upper Thames River, Maitland Valley and St. Clair Region CAs as well.

Financial constraints tend to be the biggest barrier to adopting BMPs in the CC sub-watershed. Also, many farmers are still not aware that implementing many BMPs such as improved manure storage can be profitable, especially in the long run. Whereas the religious-cultural mindscapes of many of the residents of CC may act to limit farmers' participation in EMSs like the EFP and the RWQP, this was not corroborated by the research findings reported in chapter 5 and 6 with farmers elsewhere and this is an indication of how context specific some of these findings are.

It does, however, reveal the strong need for environmental extension education in some locales. In the relative absence of publicly funded extension services by OMAFRA, environmental extension education is increasingly dependent on the conservation authorities and Ministry of Natural Resources personnel, as well as the OSCIA.

Ontario's non-farm rural residents who live within CC tend to feel that existing programs such as the NMA, the RWQP and the EFP are beneficial tools to help farmers understand and implement conservation measures. These non-farm rural people recognize, however, the importance of incentives and consideration of social characteristics in the area. They also noted that because the Canagagigue Creek does have such poor water quality, it is important that players from a variety of sectors including the industrial polluters in the area participate in mitigating the effects of pollution.

The Ausable Bayfield watershed, which drains into Lake Huron, is the focus of the fifth chapter. In this chapter, factors that motivated farmers to adopt BMPs in a relatively healthy sub-watershed were once again contrasted with the behavior of farmers in a sub-watershed with poorer quality water. The formation of a cooperative working relationship with the OFA representative in Lambton Shores/Warwick County was identified as a significant factor regarding the delivery of environmental stewardship programs. Promoting collaboration between Conservation Authorities and local farm organizations can strongly enhance the value of such programs in the eyes of the farm community and strengthen network connections between farmer organizations and environmental agencies promoting EMS.

Some types of BMPs had been done at a net cost and other BMPs like improved manure storage actually makes farms more profitable in the long run. Ausable Bayfield's farmers generally felt that small farmers were more likely to implement environmental BMPs whereas the larger scale farmers were more likely to be able to implement BMPs that also had a commercially beneficial effect. Some farmers were concerned because they did not feel that they could do everything they believed they should be doing to protect the water and the rest of the environment. Those participating extensively in outdoor recreation were often more aware of environmental issues in the Ausable Bayfield area and the importance of introducing environmental BMPs than were those not as engaged recreationally with the outdoor environment.

Interviews and focus groups with farmers, agricultural federation people and CA officials led us to the conclusion that rural conservation agents/extensionists generally understand the rural area within which they are working as well as the differing needs of larger and smaller farms, livestock, crop and mixed producers. They also know what programs and regulations they are delivering, as well as, providing proper technical support. They prefer environmental regulations that are simple and understandable. Integrated watershed planning, where all stakeholders, landowners, municipalities and ex-urbanite rural people work together to protect the environment has now become a mandatory reality but there just are not enough extensionists nor economic support for especially small farm operators to facilitate this work.

Chapter 6 describes the results of an analysis of a random sample of 317 farmers from Ausable Bayfield, the Grand River and Lake Simcoe Watersheds (worst and best sub-watersheds within each watershed). Within the Lake Simcoe Watershed, we began by determining the most and least degraded sub-watersheds in each region. Social profiles showed that the Beaver River, in Lake Simcoe watershed, contains larger, more intensive farms focusing primarily on livestock production. The Black River had considerably less agriculture and was less degraded. In the Ausable Bayfield Watershed on the other hand, the two sub-watersheds holding the most and least degraded status are comparatively similar in terms of farm size. However, Usborne River, the least degraded area, is also less dependent on agriculture and the fewer numbers who do farm usually have off-farm employment.

The Grand River Watershed includes the Upper Nith and Eramosa-Speed, considered the second most[4] and the least degraded sub watersheds respectively. Similar to the Lake Simcoe Watershed, Upper Nith possesses a greater number of farms, as well as larger farms. The type of agriculture however (livestock versus crop production) is relatively equally distributed amongst both areas.

As mentioned, the EFP was at least partially implemented by 64% of the sample. Gross farm sales, size of the farm, implementation of an EFP and an NMP, being a full time farmer and a large livestock producer were all quite significantly related to the BMP adoption rate (ARI). Sub-watershed quality, proximity to cities, and/or Conservation Authorities, age and education did not affect the adoption rate significantly.

Logistic regression models for adoption rate and gross sales helped identify the variables that predicted the dependant variables {Adoption Rate Index (ARI) and Gross Farm Sales (GFS)}. Farm size had a significant influence on both adoption rates and gross sales, the only variable which was significant in the ARI model. Thus, farmers with good management practices (higher adoption rates) and large farm sizes are likely to have better sales. The supply managed commodities (dairy and poultry) had the highest adoption rates of all commodities. However, these watersheds have many small operation farmers and this explains why the overall adoption rate is low at 0.27

---

[4] As mentioned earlier, the water quality in Canagagigue Creek (CC) is actually worse than that of the Upper Nith but the fact that Wells (2005) had conducted many interviews with CC farmers in 2005 meant that we chose instead to collect data from the Upper Nith.

out of 1.0. Once again it was clear that environmental programs have to find better ways to influence small farm operation farmers to adopt BMPs as financial and technical adoption incentives are perceived to be inadequate.

Ninety-one percent of the farmers in Ausable Bayfield, Grand River and Lake Simcoe watersheds strongly favoured environmental goods and services (EGS) payments but probably would not if it meant facing periodic environmental audits (cross-compliance) in return for the payments. Nevertheless, the recognition of the multifunctionality of private land and the impact of that land on common property within watersheds is growing as is the case for farm payments for producing EGS though the public may not accept this without cross-compliance with at least minimal regulations.

The fact that farmers most likely to adopt BMPs have higher gross farm sales and relatively large farms contrasts sharply with the prevailing view that large so-called 'factory farms' are most responsible for watershed environmental problems. Certainly intensive agricultural usage is bad for the environment but as we have seen in chapter 4, this intensity is often due to the prevalence of a large number of small farms which collectively produce more nutrients than watersheds can absorb. Of secondary importance with respect to the adoption of BMPs were the implementation of an Environmental Farm Plan, having a Nutrient Management Plan, being a full time farmer and operating a livestock enterprise. Dairy farmers are the most likely to have implemented BMPs of all commodity groups despite the fact that total farm size benefits dairying considerably less than do some other commodities like pork production (Pfeiffer and Filson, 2004). Dairy and poultry farming in Canada are supply managed enabling many of them to be able to afford to do environmentally beneficial management practices. Of course dairy producers are often heavy water users, however, and to suggest that these farms are not environmentally problematic in some ways would be wrong but dairy farmers are among the most proactive in dealing with their environmental challenges at least in part because they tend to be better financed than most other farmers in Ontario.

With the exception of dairying, off-farm employment now exceeds net farm income for farmers of all other commodities. Off-farm employment is particularly prevalent among grains and oilseeds and beef sectors where small farms predominate (Sparling, 2007). Another reason why livestock producers often implement more BMPs is that livestock farmers face a greater risk of producing excess nutrients including greenhouse gases than crop producers and the larger their scale of operation the more BMPs they need to adopt.

The contradiction between farmers and non-farm ruralresidents is a visible consequence of the way many non-farm rural people have responded to what they often perceive to be the increased environmental risks accompanying the intensification of agriculture. Public concern about water quantity and quality and environmental health has fuelled a debate about what practices are most conducive to promoting sustainable development. Governments at federal and provincial levels have promoted a stewardship driven approach but farmers' tradition of self-regulation with respect to the environment has been threatened by new regulations in the first decade of the 21st century.

DESIRABLE POLICIES FOR SUSTAINABLE AGRICULTURAL PRODUCTION

Most Western European countries have now shifted away from income support tied to production to systems where producers receive payments for the provision of EGS. Strong sustainability would protect small producers through support for multi-functional agriculture and farm payments for those producing EGSs instead of for food production (like the Canadian Agriculture Income Stabilization {CAIS} Program). In Canada a more inclusive program is needed than CAIS that will take into consideration the viability of farm business, the potentially more multifunctional nature of agriculture and the support for farmers who consider animal welfare, food safety and environmental management in their practices.

Farmers' mindscapes and in turn landscapes are shaped by their class location as worker-farmers, small capitalists and increasingly industrial capitalists. They are also shaped by their operational conditions, their norms, values, beliefs and connections with other farmers whether by kinship or farm organization. Thus, the rate of adoption of best management practices (BMPs) by farmers can be increased through interaction with adopters in their farmers' group. While farmers possess a natural affinity toward environmentally friendly behaviour because they have an obvious interest in passing on environmentally secure, viable farms to their offspring, their tight economic conditions can lead them to the point where they take short-cut management decisions that do not protect the environment. In the latter respect, somewhat surprisingly, capitalist farmers often have the wherewithal to do the right thing environmentally, despite other features of their mass production including excessive fossil fuel dependence (in contrast for example to the smaller operator Old Order Mennonites' continued use of draft animals and general avoidance of fossil fuels).

Farmers' mindscapes impact their perception of themselves and their relation to their landscapes including their environmental management and this contributes to either far or near-sighted management. Unfortunately existing federal government policy has not functioned to redress inequality among these various farming classes. As Stevens of the Ontario Christian Farmers' Organization, correctly observes:

"The current support system does not seem to support small and medium scale producers as large operations seem to walk away with the lion's share of the funding. Since support is still relative to scale, smaller operations receive very little in the form of payment. Instead, they turn to off-farm jobs and other sources of income to support their lives on the farm"(2007: 1).

As has been shown, this often jeopardizes the smaller farmers' ability to fully implement an Environmental Farm Plan, develop a Nutrient Management Plan and participate fully in environmental incentive programs like the GRCA's Rural Water Quality Program.

The policy challenge is to identify and implement an effective land stewardship mechanism that helps farmers who engage in practices that increase the production of EGS (Rae and Beale, 2008) while encouraging them to adequately take care of farm animals and comply with food safety issues like Hazard Analysis Critical Control Points (HACCP) in their production. This is extremely important if Canada wants to create an environmental niche for its farmers and keep them in production in this globally competitive market. On the one hand, the farmers should be ready for occasional, random environmental audits and compliance because the success of any program depends on cooperation and trust between the farmers and the government.

It is also important that farmer groups be created and supported (technically and financially) because our evidence suggests that farmers adopt BMPs first and foremost as the result of interaction with other farmers. Farmers' participation should also be encouraged in agri-environmental policy development but there must be room for municipalities, rural non-farm and others directly affected by these policies to have input as well. The GRCA's RWQP provides an ideal for which to strive. Despite the chasms that presently exist between many urban and rural people, as well as nonfarm rural and farmers, the transition to a more multifunctional form of agriculture can only be accomplished with support from many urban, nonfarm rural people and farmers. For this to happen, it will have to be championed by one or more political parties with a chance to affect the larger power structures.

CONCLUDING REMARKS

Water, land and air are common property resources but much of the land is privately owned so some contradictions exist between public and private interests. The tension between public versus private ownership of land impacts how we view voluntary land and water stewardship like the EFP versus publicly regulated stewardship like the NMA and CWA.

It is precisely these contradictions within, between and among the various systems affecting the production, processing and distribution of food that are key to understanding where our evolving food regime is going and what environmental management systems are most appropriate for farmers. Our research has shown that we need a voluntary approach to environmental management backstopped by regulatory control, supported by financial and technical incentives with continuous collaboration between landowners and municipal, provincial and federal governments.

Most farmers' present mindscapes and conditions are such that they experience a personal feeling of responsibility for protecting the environment, yet tend to believe that environmental regulation is excessively intrusive and severely limiting. This has happened in part because some regulations were brought in without actively involving farmers sufficiently from the beginning thus engendering mistrust and resentment towards both government officials and the imposed regulations.

The time has come in Ontario to consider developing cross-compliant environmental management systems with farmers. Cross-compliant systems provide an alternative approach to the present practice in Ontario and the environmental advantages and the prospect of paying farmers for their production of EGSs could benefit Ontario without costing the public too much more than they are presently paying for environmental management. Such a program could be optional for successful producers with more than one quarter of a million dollars in gross farm sales though differential targeting of payments for producing EGSs may be politically impossible to achieve and therefore unwise.[5]

---

[5] In the EU, extension workers (not private consultants) help to ensure that farmers have nutrient management plans and appropriate phosphorus, nitrogen and land to animal balances. In return for significant single farm payments farmers allow semi-open access to the public via hiking trails, school visits to their farms and regular adoption of such BMPs. These include protecting wetland areas for wildlife, set asides, riparian buffer strips, cattle fenced off from streams, no-till cultivation, contour farming around hills and no application of manure in the winter due to the run-off dangers.

As the number of farms continues to decline, more intensive, larger operation farms are expanding and this threatens the continued existence of small family farms. Clearly, the latter will need different strategies to survive than the larger farms whose operations are usually more profitable. More than half of Canada's farms are too small to be viable and a sizable percentage of those with revenues of less than $250,000 are in trouble (Sparling and Laughland. 2006). Most small farming operations' only real prospect for survival is via off-farm employment. However, receiving increased payments for producing EGS does present the prospect that many of these now marginal farms could survive over at least the short-term.

Different policies are therefore required for the different sectors, each of which needs to develop strategic plans unique to their sector. The next Agricultural Policy Framework and a revised Canadian Agricultural Income Stabilization policy could emphasize the environmental branding of Canadian agriculture only if a substantial shift toward support for cross-compliant EGS payments occurred. This would be an explicitly more sustainable rural community approach minimizing outward migration of farm youth while enabling continuing efficiency enhanced productivity improvements.

In addition to securing food security, multifunctional agriculture is about conserving the land, protecting the natural environment, developing satisfying landscapes, maintaining people's cultural heritage, providing recreational opportunities and the viability of rural communities. In order to achieve a truly greater environmental security, a multifunctional agriculture combined with a collaborative approach to watershed management is required.

Intensive agricultural production reduces biodiversity, contributes to climate changing greenhouse gas production, depends heavily on non-renewable energy sources, generates excess nutrients and also uses many chemical forms of fertilizer and plant protection. Yet, of course, it provides us with a bountiful harvest of food and fibre and contributes to the provision of some EGS so long as appropriate BMPs are implemented. Though some of these BMPs are done voluntarily, with changing demographics in rural areas, and concerns for environmental and human health alike, governments have received pressure to regulate agriculture more aggressively so that environmental standards can be better safeguarded.

A stronger form of sustainability will therefore likely only be developed when government departments overseeing agriculture, conservation authorities and farm organizations adopt structures which will enable a multifunctional form of agriculture wherein farmers are rewarded not only for their food and fibre production but for their production of environmental goods and services.

If Ontario farmers accept the need to behave more multifunctionally they ought then to be compensated with existing available resources, simply distributed differently, for producing EGS. Not only will this be good for the environment and the general public, this may be the only hope for small operators' survival.

Fortunately there are many counter tendencies to the globalizing food regimes, the internationalization of junk food and the constant pressures "to get big or get out". The rising influence of the locavore and slow food movements including the quest for more locally grown ethnoculturally diverse vegetables and fruits, the growing interest in organic agriculture, the rise of community shared agriculture and farmers' markets all herald a sea change in reaction to the latest food regime and the dominance of multinational agri-business. This is a struggle for healthy ecosystems and healthy food and it can still be won!

GLOSSARY[6]

Agriculture and Agri-Food Canada: "Agriculture and Agri-food Canada provide information, research and technology, and policies and programs to achieve security of the food system, health of the environment and innovation for growth".[7]

Agricultural Institute of Canada (AIC): The AIC represents a group of individuals and organizations involved in Canadian agriculture, food, environment and health. It has the power to influence national public policy, it informs the public on issues and possible science-based solutions, and strives to achieve enhanced sustainable communities and food systems.[8]

Agricultural Intensification: Rising levels of capitalization through purchased non-farm inputs combined with rising outputs per hectare. *

Agri-tourism: An aspect of landscape multifunctionality where farmers link travel with agricultural products, services or experiences. Farmers are able

---

[6] Prepared by Melissa Yule.

* All definitions marked with an asterisk are taken from G.C. Filson. 2004. Intensive Agriculture and Sustainability: A Farming System's Analysis. Vancouver-Toronto: UBC Press.

[7] Agriculture and Agri-Food Canada, Government of Canada "About Agriculture and Agri-Food Canada", online at <http://www.agr.gc.ca/aafc_e.phtml> (retrieved November 6, 2006).

[8] Agricultural Institute of Canada, "About AIC," online at <http://www.aic.ca/about/index.cfm> (retrieved November 25, 2006).

to supplement their income with the earnings of camping facilities or Bed and Breakfasts and the use of their land for skiing, hiking or other recreational activities.

Agro-climatic ratings: Agro-climatic ratings are based on several indices created by the Agriculture Canada Research Station . Examples of indices include heat units, seasonal water deficits and aridity. Indices were determined for the main agricultural regions of Ontario and Quebec for the baseline (observed) climate (1961- 1990 average) and for three future time periods (2010-2039, 2040-2069, 2070-2099). All of Canada's top agro-climatic ratings (Agro-Climatic Rating Index, rating 2-3) are in southern Ontario (Hilts, 2006).

Agro-ecosystem landscapes: Landscapes where agricultural production and ecosystem functions are occurring simultaneously. In agro-ecosystems, both components are highly dependent on, and strongly influenced by, each other. The environmental security of the agro-ecosystem is an important consideration in order to improve the sustainability of Ontario rural communities where agriculture is practiced. The major feature of a resilient agro-ecosystem system is the capacity farmers have to perceive options, and to be able to choose those options that will promote environmental security rather than undermine it.

Anthropocentrism: Human centeredness, where humans are the main concern.

Best Management Practices: Tools and techniques that are designed to restore and protect the environment. They may include, for example, the use of riparian buffer strips between crops and streams, or variable input technologies that ensure excess nutrients are not applied to crops. *

Biocentrism: As the antonym of anthropocentrism, biocentrism is the belief that all life is of equal value and importance.

Biodynamic farming: A method of farming that aims to not only protect the environment, but to enhance all ecosystem processes. Farms are self-contained entities where crop rotation, composting, inter-planting and careful treatment of livestock are practiced. In addition, the influences from the moon and other planets are also taken into account. [9]

British North American (BNA) Act: The British law creating Canada in 1867. It is now known as the Constitution Act of 1867.

---

[9] Society for Biodynamic Farming and Gardening in Ontario, "What is Biodynamic Farming?", online at <http://www.biodynamics.on.ca/whatisbd.htm> (retrieved November 7, 2006).

Canadian Environmental Protection Act: "An Act respecting pollution prevention and the protection of the environment and human health in order to contribute to sustainable development." [10]

Cash crop: Crops grown for sale rather than subsistence.

Chlorobenzene: Chlorobenzene is a colorless, flammable liquid that does not occur naturally in the environment. It is used as a solvent for some pesticide formulations. If released to the air, chlorobenzene is slowly broken down by reactions with other chemicals and sunlight or can be removed by rain. In water, it will rapidly evaporate to the air and/or be broken down by bacteria. When released to soil, it is broken down rapidly by bacteria, but some will evaporate to the air and some may filter into the groundwater. Chlorobenzene does not build up in the food chain. [11]

Christian Farmers' Federation of Ontario (CFFO): The goals of the CFFO include enabling farmers to work out their Christian faith in their vocation as citizens, and to develop public policy applications on major issues affecting Ontario agriculture based in the Christian faith. [12]

Classification of soils: There are two components in the classification of soils for capability rating. These consist of the capability classes and the capability subclasses. Together they provide information on the degree and kind of soil limitation(s) present in the area of question and indicate the general sustainability of the soil for the agricultural use in question. The classes are as follows:

Class 1: Soils in this class have no significant limitations in their use for common cultivated crops.

Class 2: Soils have a moderate limitation, or a combination of minor limitations which restrict the range of crops they can support or require moderate conservation practices.

Class 3: Soil has a major limitation or combination of lesser limitations which restrict the range of crops or require special conservation practices. [13]

---

[10] Government of Canada, "Canadian Environmental Protection Act, 1999, c. 33," online at <http://laws.justice.gc.ca/en/C-15.31/225697.html> (retrieved November 15, 2006).

[11] Agency for Toxic Substances and Disease Registry (ATSDR) "ToxFAQs: Chlorobenzene," online at www.atsdr.cdc.gov/tfacts131.html (retrieved November 25, 2006).

[12] Christian Farmers Federation of Ontario (CFFO) "About CFFO," online at <www.christianfarmers.org/main_about_cffo/main_about_affo.html> (retrieved November 15, 2006).

[13] L.W. Schut, and E.A. Wilson, "Agricultural Capability Ratings and System of Classification," online at <http://sis.agr.ca/cansis/publications/on/on-84-3/on-84_report.pdf> (retrieved November 18, 2006).

This book mentions only classes 1-3 which are explained here. There are a total of 7 classes and many subclasses which can be found at http://sis.agr.ca/cansis/publications/on/on-84-3/on-84_report.pdf.

Clean Air Act: Canada's Clean Air Act aims to reduce emissions of both air pollutants and greenhouses gases and protect human health and the environment. [14]

Community Shared Agriculture (CSA): The CSA movement grew out of biodynamic farming as a system of growing and distributing organic produce that was designed to restore the traditional producer - consumer relationship. Non-farmers purchase a share of a farm at the beginning of the year's harvest and receive fresh, wholesome produce throughout the season. In return, farmers receive the start-up capital necessary for purchasing seeds, supplies and soil amendments. [15]

Conservation tillage: Conservation tillage is a soil conservation method that allows crops to be grown with minimal cultivation of the soil (PEET 2001). *

Cost Sharing: Cost-sharing is a practice widely applied as an incentive for program adoption where program costs are shared by the producer and the government body or agency that administers the program. To qualify landowners must first apply for project funding and perform the practices according to the specifications of the management plan before they can receive payment. The revised NMA can provide through the Federal Government cost sharing up to $30,000 split on a 30 to 50% basis. Cost-sharing between the federal and provincial governments provides up to 90% of funding for manure storages so farmers can get up to $60,000 for manure storage and handling. Although the financial support is significant, farmers may not have the start up capital required and choose not to participate.

Cross-compliant systems: Systems where all government payments to farmers must be accompanied by compliance with environmental regulations, in exchange for the delivery of funds including tax concessions to farmers which support environmentally beneficial management practices (BMPs) (Shukla,. 2005) If farmers fail to do meet regulations their governmental funding is cancelled, potentially forcing some farmers to sell livestock or

[14] Environment Canada, Government of Canada, "Canada's Clean Air Act," online at <http://www.ec.gc.ca/cleanair-airpur/Clean_Air_Act-WS1CA709C8-1_En.htm> (retrieved November 30, 2006).
[15] Green Venture, "Community Shared Agriculture," online at < www.greenventure.ca /gv.asp?ID+123> (retrieved November 18, 2006).

take land out of production. These regulations, for instance, set out rules for nitrogen and phosphorus application in return for which the farmers receive funding per hectare.

Diffusion of Innovation: Farmers who are the quickest to adopt conservation measures may persuade the more pessimistic farmers to become environmental stewards by demonstrating the benefits of sustainable practices. Attitudes and beliefs greatly affect a farmer's operational decisions, and observing voluntary adoption within their own peer group, and seeing success, increases the viability of environmental programs amongst the more pessimistic farmers.

Dichloro-diphenyl-trichloroethane (DDT): DDT is a pesticide that came into use in the 1940s. It was widely employed to control insects that were either damaging to crops or known to carry diseases such as malaria and typhus. DDT is a white, crystalline solid with no odour or taste. It binds strongly to soil and is broken down slowly by micro organisms, in some cases taking 2-15 years to disintegrate. DDT builds up in plants and in fatty tissues of fish, birds, and other animals, and it leads to the thinning of bird's egg shells and may be linked to cancer. Due to its damaging effects on wildlife, in 1972 the use of DDT was banned in the United States but it continues to be used in many countries to combat insect born diseases. [16]

Dioxin: Dioxin is a name typically used when referring to halogenated organic compounds. They are the by-products of various industrial processes (i.e., bleaching paper pulp, and chemical and pesticide manufacture) and combustion activities (i.e., burning household trash, forest fires, and waste incineration), although it has no known use. Dioxins are found at low levels throughout the world in air, soil, water, sediment (the bottom of rivers, streams, and lakes), and in foods such as meats, dairy, fish, and shellfish. The highest levels of dioxins are usually found in soil, sediment, and in the fatty tissues of animals due to lipophilic properties. [17]

Economic corridor: A path along which goods, and therefore money, can flow. In most cases economic corridors are referring to roads, waterways and railways.

---

[16] Agency for Toxic Substance and Disease Registry (ATSDR) "ToxFAQs: for DDT, DDE, and DDD," online at <http://www.atsdr.cdc.gov/tfacts35.html> (retrieved November 25, 2006).

[17] Agency for Toxic Substances and Disease Registry (ATSDR) / Division of Toxicology and Environmental Medicine (DTEM), "ToxFAQs: CABS, Chemical Agent Briefing Sheet: Dioxins," online at <http://www.atsdr.cdc.gov/cabs/dioxins/dioxins_cabs.pdf> (retrieved November 25, 2006).

Economies of scale: A company that achieves economies of scale lowers the average cost per unit through increased production since fixed costs are shared over an increased number of goods. [18]

Ecosystem perspective: A systems approach that considers the human component as an intrinsic part of the ecosystem. It allows for a holistic view of the system, as well as the interactions its different components.

Environmental agricultural extension: The aim of agricultural extension is to promote the reflection on the impact of our actions on the environment, and to consider alternative interaction processes. Environmental extension processes can guide policy development and help re-structure institutions involved with environmental management and research. The studies found in this book are concerned with individual daily decisions, farmers' responsibility for, and motivations behind, the changes they effect on the environment in the process of food production.

Environmental Farm Plan (EFP): The Farm Plan refers to documents that are voluntarily prepared by farm facilities to raise their awareness of the environment on their farm. "Through the EFP process, farmers will highlight environmental strengths on their farm; identify areas of environmental concern and set realistic goals and timetables to improve environmental conditions" (OMAF 2003d, 1). *

Environmental Goods and Services: Described as the environmental benefits that result from sustainable production, environmental goods and services are a good news story. Although frequently equated to carbon credits, environmental goods include clean air, clean water, biodiversity, and technologies that reduce greenhouse gas emissions. Environmental services include renewable energy, environmental education, training and information, and waste and remediation of soil, surface water, seawater and groundwater, and scenic vistas.[19]

Environmental Management Systems (EMS): Environmental Management Systems are a framework created by an individual organization to improve its environmental performance. The development and implementation of EMS help to ensure departments and agencies meet legislation and policy

[18] Investopedia, "Economies of Scale," online at <www.investopedia.com/terms /e/economiesofscale.asp> (retrieved November 29, 2006).
[19] Statistics Canada, "Environment Industry: business sector," The Daily, online at <http://www.statcan.ca/Daily/English/040921/d040921c.htm> (retrieved November 29, 2006).

objectives, and demonstrate due diligence when making decisions and managing risks. [20]

Environmental security: Environmental security is related to the degree of environmental threats to human life (Nef, 1999; Falkenmark, 2001; Klubnikin and Causey, 2003). The complexity of ecosystems' structure and processes, as well as their intrinsic uncertainty, only increase the challenge to find effective approaches to enhance environmental security. In the agricultural context, there is concern about the high dependence of, and the strong influence on, environmental conditions. For these reasons it is imperative that farmers view their land as agro-ecosystems and implement management programs that will ensure their sustainability.

Equivalent Agri-Environmental Plan (EAEP): EAEP has similar characteristics to the Environmental Farm Plan in that is carried out by an organized group of farmers; it is based on physical boundaries such as watershed, aquifers, landscape; it addresses a single high priority issue and strives to achieve cumulative benefits.[21] The EAEP differs from the EFP, however, in that it is a group farm plan which selects a single issue to influence management beyond each individual's 'own farm', and promote the strategic targeting of issues so that resources can be allocated to deal with those issues. At this time, Ontario has chosen not to participate in the EAEPs (Chekay, 2006).

Escherichia coli: E. coli is a bacterium often found in the gut of humans and other warm blooded animals. Most strains are harmless while others can cause abdominal pain, diarrhoea and even death. In May 2000 ,Walkerton Ontario's town water was found to contain the E. coli strain O157:H7. An improperly operated well near a beef farm failed to detect or report the contamination, and consequently 2300 people became infected and 7 died. [22]

Eutrophication: In aquatic ecosystems, high nutrient concentrations stimulate algal blooms that can block sunlight and inhibit the growth of all plants. As the dying vegetation decays, it consumes most of the dissolved oxygen which is needed by all respiring aquatic life. Although eutrophication is a

---

[20] Government of Canada, "Environmental Management Systems," online at <http://www.greeninggovernment.gc.ca/Default.asp?lang=En&n=A4FA4E9C-1> (retrieved November 21, 2006).

[21] Agriculture and Agri-food Canada, Government of Canada, "Canada's Agriculture Policy Framework, Environmental Stewardship Programs," online at < http://www.aic.ca/conferences/pdf/2005/Dean_Smith_ENG.pdf> (retrieved November 25, 2006).

[22] World Health Organization (WHO), "Enterohaemorrhagic Escherichia coli (EHEC)," online at <http://www.who.int/mediacentre/factsheets/fs125/en> (retrieved November 7, 2006).

natural process, human activities (with agriculture being a main contributor) increase the rate at which nutrients enter aquatic ecosystems. Run-off from agricultural land in the surrounding watershed carries vast amounts of nitrogen, phosphorus and other organic compounds to bodies of water, disrupting natural cycles and gravely impacting marine life.

Externalities: Externality is a broad term that includes all actions of producers or consumers that have unintended external effects on other producers and/or consumers. These effects are considered residuals, intangibles, and incommensurables, all factors that aren't usually accounted for in cost calculation. Externalities can be either positive (benefiting the external party), as is the case with a technological spill over where not only the inventor benefits, but the rest of society does as the technology enters the general pool of technological knowledge. Externalities can also be negative (harmful to the other party), for instance, pollution. When a factory discharges its untreated effluents in a river, the river is polluted and consumers of the river water bear costs in the form of health costs or/and water purification costs. [23]

Full cost accounting (FCA): FCA generally refers to the process of collecting and presenting information (costs as well as advantages) to decision makers on the trade-offs of each proposed alternative. FCA focuses on three major types of costs: up-front costs, operating costs, and back-end costs. Other categories of costs that can be included, but require special consideration, are remediation costs at inactive sites, contingent costs, environmental costs, and social costs. [24]

Genetically Modified crops: Crops whose' genes have been manipulated in order to favour existing traits, or promote the expression of new traits, that will prove beneficial for the plant or consumers.

Greenhouse gases: Emissions that include carbon dioxide, methane, and nitrous oxide, among others. These gases prevent the sun's ultraviolet rays from being reflected back into the atmosphere, causing the earth's surface to warm. *

Hazard Analysis and Critical Control Point (HACCP): HACCP is a tool that can be useful in the prevention of food safety hazards (biological, chemical, and physical) which may pose an unacceptable health risk to the

---

[23] U. Sankar, "Environmental Externalities," online at <http://coe.mse.ac.in/dp/envt-ext-sankar.pdf> (retrieved November 29, 2006).

[24] Environmental Protection Agency, Government of United States of America, "Full Cost Accounting," online at <http://www.epa.gov/epaoswer/non-hw/muncpl/fullcost/wh atis.htm> (retrieved November 18, 2006).

consumer. While extremely important, HACCP is only one part of a multi-component food safety system that includes food manufacturing practices, sanitation standard operating procedures, and a personal hygiene program. [25]

Hydro modification: The U.S. Environmental Protection Agency defines hydro modification as the "alteration of the hydrologic characteristics of surface waters, which in turn could cause degradation of water resources." The construction of dams, tide gates, culverts, bridges, piers, and jetties, as well as the armouring of shorelines and the placement of fill, have helped create drinking water supplies, reduce flood impacts, expand road networks, improve navigation, increase drainage, prevent erosion, and reduce sediment loss. Many of these activities have also led directly or indirectly to adverse impacts on aquatic ecosystems. [26]

Indicators of sustainability: The National Agri-Environmental Health Analysis and Reporting Program (NAHARP) developed indicators of water, land, air and biodiversity quality using scientifically based models in part through the use of the Farm Environmental Management Survey of 2001 (Jayasinghe-Mudalige et al., 2005). Agriculture and Agri-food Canada (AAFC) has been working with the International Institute for Sustainable Development (IISD) and a Social Indicator Technical Working Group which includes farmers, government and nongovernmental personnel as well as academics to develop socio-economic indicators. This group's indicators include, among others, farm family finances, perceived qualify of life on the farm, socio-economic infrastructure, formal/informal governance, empowerment through collective organization and farm family succession.

Integrated Pest Management (IPM): IPM is a program designed to help manage diseases, insects, weeds and animal pests through pest management recommendations that minimize the use of chemical pesticides (OMAF 2003b). *

International Institute for Sustainable Development (IISD): In 1988, Brian Mulroney announced that Canada was prepared to launch and fund a sustainable development institute in response to the World Commission on the Environment. The IISD was created in Winnipeg in 1990 and

---

[25] J.E. Rushing and D.R. Ward, NC State University College of Agricultural and Life Sciences, "HACCP Principles," online at <http://www.ces.ncsu.edu/depts/foodsci /ext/pubs/ haccpprinciples.html> (retrieved November 25, 2006).

[26] Washington State Department of Ecology, "Hydro modification," online at <http://www.ecy.wa.gov/pubs/9926/5f.pdf> (retrieved November 25, 2006).

funded wholly by the governments of Canada and Manitoba. Through research and effective communication of their findings, the IISD "engages decision-makers in government, business, NGOs and other sectors to develop and implement policies that are simultaneously beneficial to the economy, the global environment and to social well-being". [27]

International Organization for Standardization (ISO): A network of the national standards institutes of 157 countries makes up this non-governmental organization. Universal standards are achieved through consensus agreements between national delegations representing all the economic stakeholders concerned - suppliers, users, government regulators and other interest groups, such as consumers. They agree on specifications and criteria to be applied consistently in the classification of materials, in the manufacture and supply of products, in testing and analysis, in terminology and in the provision of services. [28]

Intensive livestock operations (ILOs): Large scale cattle, hog and poultry farming where manure and waste disposal are controlled. Liquid manure is stored in lagoons until it is spread on the fields as a fertilizer. *

Kyoto Protocol: An agreement was struck among many countries to accept treaty rules and cut back on their emissions of carbon dioxide and other global warming gases to a level of 6% below their 1990 levels by the year 2010. The Kyoto Protocol has yet to be accepted by all countries, and yet to be implemented in many countries that signed the protocol, including Canada. *

Landscape multifunctionality: The concept of landscape multifunctionality recognizes the multiple roles that landscape plays including meeting ecological, socio-cultural and society's production needs. Ecological functions involve support for biodiversity, natural habitat and the recharging of groundwater; social functions include recreation, cultural heritage, aesthetics and regional identity; and the production functions range from forestry to agriculture, hunting and water consumption (Brandt and Vejre, 2004; Wilton, 2005).

Low phosphorus pig: To address the problem of manure-based environmental pollution in the pork industry, scientists at the University of Guelph have developed the phytase transgenic pig. The saliva of these pigs contains the

---

[27] International Institute for Sustainable Development (IISD), "FAQ," online at <http://www.iisd.org/about/faq.asp> (retrieved November 25, 2006).

[28] International Organization for Standardization, "Overview of the ISO System," online at <http://www.iso.org/iso/en/aboutiso/introduction/index.html> (retrieved November 25, 2006).

enzyme phytase, which allows the pigs to digest the phosphorus in phytate, the most abundant source of phosphorus in the pig diet. Without this enzyme, phytate phosphorus passes undigested into manure to become the single most important manure pollutant of pork production.[29]

Mindscape: The mindscape is a complex structure that enables perception and cognition processes, which are at the basis of our decision making process. It is defined by a combination of different elements, including perception, knowledge (encompassing conscientization and awareness), worldview, and values (priorities, expectations, preferences and motivations) (Marzall 2006).

Mixed-wood Plains region: The Mixed-wood Plains region lies along a narrow plain in the southern parts of Ontario and Quebec. The region's fertile soil, hot summers, and abundant water supply provide ideal conditions for ample livestock and agricultural production. Roughly 37 percent of Canada's agricultural production comes from this ecozone even though it occupies only 9 percent of the country's landmass and has a population density is over 100 persons per square kilometre — 10 times higher than anywhere else in Canada. [30]

National Agri-Environmental Health Analysis and Reporting Program (NAHARP): The NAHARP came into being shortly after Agriculture and Agri-food Canada created the agri-environmental indicators (AEIs) to strengthen the departmental capacity in the development and continuous improvement of AEIs.[31]

National Farm Stewardship Program: National Farm Stewardship Program (NFSP) provides technical and financial assistance to support the adoption of beneficial management practices (BMP) by agricultural producers and land managers. Producers who have a completed and reviewed environmental farm plan (EFP) or equivalent agri-environmental plan are eligible to apply for financial and technical assistance to implement the

[29] S.P. Golovan, G.R. Meidinger, A. Ajakaiye, M. Cottrill, M.Z. Wiederkehr, D.J. Barney, C. Plante, J.W. Pollard, M.Z. Fan, M.A. Hayes, J. Laursen, J.P. Hjorth, R.R. Hacker, J.P. Phillips, and C. W. Forsberg. 2001. "Pigs expressing salivary phytase produce low-phosphorus manure," Nature Biotechnology, 19: 741-745 online at <http://www.nature.ca/nbt/journal/v19 /n8/abs/nbt0801_741.html> (retrieved November 25, 2006).

[30] Canadian Geographic, "Mixed-wood Plains," online at <http://www.canadiangeographic.ca /atlas/themes.aspx?id=mixedwood&sub=mixedwood_industry> (retrieved November 25, 2006).

[31] Agriculture and Agri-Food Canada, Government of Canada, "National Agri-Environmental Health Analysis and Reporting Program (NAHARP)," online at <http://www.agr.gc.ca/env /naharp-pnarsa/pdf/naharp-pnarsa_e.pdf> (retrieved November 29, 2006).

BMPs identified in their EFP. Approved applicants are eligible for a maximum of $50,000 in federal funding. [32]

Nitrate: Nitric acid can form salts which are called nitrates. These salts are limiting factors in plant growth since they are easily lost through leaching and de-nitrification by bacteria. Nitrates are therefore a main component of fertilizers and they have become an environmental issue since they are carried off by run-off and collect in water sources where they increase the rate of eutrophication. [33]

n-Nitrosodimethylamine: n-Nitrosodimethylamine is a yellow liquid that is formed during various manufacturing processes in air, water, and soil from reactions involving other chemicals called alkylamines. It is also found in some foods and may be formed in the body. When released to the air, it is broken down by sunlight in a matter of minutes. In water it may break down when exposed to sunlight or by natural biological processes and when it is released to soil, it may evaporate into air or sink down into deeper soil. [34]

Non-point source (NPS) pollution: Pollution that is the result of many diffuse sources as opposed to originating from one location. An example of NPS pollution is when runoff gradually picks up, and carries away, natural and human-made pollutants, which then are deposited into lakes, rivers, wetlands, coastal waters, and our underground sources of drinking water.

Nutrient Management Act (NMA): An Ontario law that sets out standards for nutrient management on farms while protecting the environment (OMAF 2003c). *

Nutrient Management Plan (NMP): The objective of NMPs is to use nutrients such as phosphorus, nitrogen and potassium in a proper manner for the best possible economic benefit, while minimizing the impact on the environment (OMAF 2003a). *

Nutrient Management units: The amount of nutrients that give the fertilizer replacement value of 43 kilograms of nitrogen or 55 kilograms of phosphate of nutrient as established by the Nutrient Management

---

[32] Agriculture and Agri-food Canada, Government of Canada, "The National Farm Stewardship Program," online at <http://www.agr.gc.ca/env/efp-pfa/index_e.php?section=nfsp-pnga&page=intro> (retrieved November 25, 2006).

[33] Wikipedia, "Nitrates," online at <http://en.wikipedia.org/wiki/Nitrates> (retrieved November 25, 2006).

[34] Agency for Toxic Substance and Disease Registry (ATSDR), "ToxFAQs for Nitrosodimethylamine," online at <http://www.atsdr.cdc.gov/tfacts141.html> (retrieved November 25, 2006).

Protocol.[35] As of July 2003, all new livestock operations and expanding operations above 300 Nutrient Management Units are required by the NMA to have Nutrient Management Plans (NMPs) and will be subject to regulation. There is also the expectation that all farms must have NMPs by 2008 (Caldwell, 2005).

Ontario Farm Environmental Coalition (OFEC): A group consisting of most major agricultural organizations in Ontario. It has the formal membership and structure of a general forum as well as a steering committee and various working groups to deal with specific issues. The OFEC was the creator of the "Our Farm Environmental Agenda" that aimed to have every Ontario farmer develop and implement an environmental plan, tailored to his/her own particular farm operation. The OFEC formed an *Environmental Farm Plan (EFP)* Working Group that sculpted the EFP model, and later the Water Quality Working Group (WQWG) was formed to focus on water quality and more recently the OFEC established a Nutrient Management Working Group to focus on developing what is now the Nutrient Management Planning (NMP) Strategy. [36]

Ontario Ministry of Food and Rural Affairs (OMAFRA): "The Ministry of Agriculture, Food and Rural Affairs helps to build a stronger agri-food sector by investing in the development and transfer of innovative technologies, retaining and attracting investment, developing markets, providing regulatory oversight, and providing effective risk management tools. The Ministry also helps enable rural Ontario to build strong, vital communities with diversified economies and healthy social and environmental climates." [37]

Ontario Farm Coalition (OFEC): The OFEC is a coalition of Ontario farm organizations led by the Ontario Federation of Agriculture (OFA), Christian Farmers Federation of Agriculture (CFFO), Agricultural Groups Concerned About Resources (AGCare), and The Ontario Farm Animal Council (OFAC). It was established in July 1991 to set a sound agenda for

---

[35] Ministry of the Environment. Government of Ontario, "Nutrient Management Act, 2002, Ontario Regulation 267/03," online at <http://www.e-laws.gov.on.ca/DBLaws/Regs/English/030267_e.htm> (retrieved November 18, 2006).

[36] Ontario Farm Environmental Coalition, Government of Canada, "Groups and Organizations in Ontario Involved in Manure Management Issues," online at <http://res2.agr.ca/initiatives/manureneten/ofec.html> (retrieved November 30,2006).

[37] Government of Ontario, "Service Ontario Info-go," online at <http://www.info go.gov.on.ca/infogo/office.do?actionType=servicedirectory&infoType=service&unitId=UN T0000319&locale=en> (retrieved November 25, 2006).

Ontario's farming community that addresses environmental issues associated with agricultural practices. [38]

Our Farm Environmental Agenda (OEFA): OEFA is a plan that highlights environmental concerns in modern agriculture from soil erosion to nutrient management, inputs, and wildlife habitat. It laid the groundwork for the Environmental Farm Plan. [39]

Ontario Soil and Crop Improvement Association (OSCIA): The OSCIA strives to communicate and facilitate responsible, economic management of soil, water, air and crops. The Association works in partnership with the provincial and federal agricultural departments to deliver both the Greencover Canada Program and the on-farm portion of the Canada-Ontario Water Supply Extension Program (Tier 1). The OSCIA is also the program delivery agent for the federal government's Ontario Farm Stewardship Program, Environmental Farm Plan programs and the Nutrient Management Financial Assistance Program. [40]

Organic farming: Organic farming produces such commodities as grain, produce, dairy and other products without the use of chemicals such as pesticides and fertilizers. *

Pathogen: A pathogen or infectious agent is a biological agent that causes disease or illness to its host. One of the primary pathways by which food or water become contaminated is from the release of untreated sewage into a drinking water supply or onto cropland, with the result being that people who eat or drink from contaminated sources become infected. The contamination of Walkerton's drinking water with E.coli bacteria is a good example of this. [41]

Pesticides: A form of chemical repellent. The word "pesticide" can be used in a very general sense to include herbicides (weeds), fungicides (fungus) and insecticides (insects).

Phosphorus: Phosphorus is often a limiting nutrient in many environments and it governs the rate of growth of many organisms. For these reasons

---

[38] Agriculture and Agri-food Canada, Government of Canada, "Ontario Farm Environmental Coalition," online at <http://res2.agr.ca/initiatives/manurenet/en/ofec.html> (retrieved November 25, 2006).

[39] AGCare, "Time to take back Our Farm Environmental Agenda," online at http://www.agcare.org/new.cfm?keyissueid=4&documentid=305 (retrieved November 25, 2006).

[40] Ontario Soil and Crop Improvement Association, "Ontario Soil and Crop Improvement Association," online at <http://www.ontariosoilcrop.org> (retrieved November 30, 2006).

[41] Wikipedia, "Pathogen," online at <http://en.wikipedia.org/wiki/Pathogen> (retrieved November 25, 2006).

phosphorus is a major component of fertilizers spread on fields to maximize crop growth. One risk involved with fertilizer application is that field run-off carries off the nutrients, including phosphorus, that typically end up in bodies of water where they accelerate eutrophication.

Phytase: "An enzyme that breaks down the indigestible phytic acid (phytate) portion in grains and oil seeds, thereby releasing digestible phosphorus and calcium for the pig." [42]

'Places to Grow' Act: An act that provides the legal framework necessary for the provincial government to designate geographic areas as growth zones and enables it to develop a growth plan in collaboration with local officials and stakeholders to ensure their particular needs are being met (Places to Grow, 2005).

Planning Act: An act that promotes sustainable economic development; provides for a land use planning system led by provincial policy; integrates matters of provincial interest in provincial and municipal planning decisions; ensures planning processes are open, accessible, timely and efficient; and recognizes the decision-making authority and accountability of municipal councils in planning. [43]

Point source of pollution: Pollution that diffuses from one particular channel, such as a pipe or a ditch, making it easier then non-point source pollution to resolve. It is typically concentrated and is the most significant contamination source.

Pollution Probe: A Canadian environmental organization that conducts extensive research into environmental problems, promotes widespread understanding through education, and aims to find practical solutions through advocacy. They work in partnership with government agencies, private businesses and other non-profit organizations that have a legitimate interest in an issue in order to find solutions. [44]

Precautionary principle (PP): A management approach where the actions taken are proactive rather than reactive, responding to a disaster after it has occurred. *

---

[42] L. McMullen and P. Holden, "Phytase fact sheet," US Department of Agriculture (USDA), online at <http://extension.agron.iastate.edu/immage/pubs/phytase.doc> (retrieved November 26, 2006).

[43] Government of Ontario, "Planning Act 1994, c. 23, s. 4," online at <http://www.e-laws.gov.on.ca/DBLaws/Statutes/English/90p13_e.htm> (retrieved November 25, 2006).

[44] Pollution Probe, "Who We Are," online at <http://www.pollutionprobe.org> (retrieved November 25, 2006).

Remote sensing: In general, "remote sensing is the measurement or acquisition of information of an object or phenomenon, by a recording device that is not in physical or intimate contact with the object". Its uses range from Earth observation and weather satellites, to monitoring of a fetus in the womb via ultrasound. [45]

Riparian buffer strips: Zones of ecological compensation (ZECs) along the margins of waterways, such as hedges with herbs and native flora, that can provide the resources needed for beneficial organisms to develop, which can help keep pests and disease in check. *

Sierra Club of Canada: The Sierra Club of Canada if a public interest group with the aspiration to advance the preservation and protection of the natural environment with charitable resources. It focuses on four main areas for its national campaigns which are: Health, biodiversity, the atmosphere and energy; and a sustainable economy. The Sierra Club has the potential to galvanize the government into action as was proven when it provided information for Justice O'Connor's report following the Walkerton disaster which resulted in the Nutrient Management Act and the Clean Water Act. [46]

Source Protection Planning (SPP): Source Protection Planning is a piece of legislation that resulted from the Walkerton Report which aims to ensure water quality in Ontario. SPP provided for in the CWA takes a collaborative, watershed approach to risk based priority setting by relevant stakeholders. .

Sustainable agriculture: Term used to refer to production systems that are environmentally benign (or enhancing), economically viable, and socially acceptable. Five system level attributes form the basis elements of sustainability: productivity, security, protection, viability, and acceptability. *

Theory of Altruism: S.H. Schwartz's (1972,1977) norm-activation theory of altruism holds that actions occur in response to personal moral norms about those actions, and that these are activated in individuals who believe that certain conditions pose threats to other people. Movements based on altruistic values emphasize the importance of collective goods where an

---

[45] Wikipedia, "Remote Sensing," online at <http://en.wikipedia.org/wiki/Remote_sensing> (retrieved, November 29, 2006).

[46] Sierra Club of Canada, "About Us," online at < http://www.sierraclub.ca > (retrieved November 25, 2006).

individual's awareness of consequences causes them to initiate actions that could avert those consequences. [47]

Watershed Approach: The watershed approach is a framework for environmental management that focuses public and private sector efforts to address the highest priority problems within hydrologically-defined geographic areas, taking into consideration both ground and surface water flow. [48] New York's watershed approach involved paying upstream farmers to provide for almost all of the environmental goods and services they produced. This has reduced New York City's water treatment costs dramatically, dropping from $8 billion to about $1.5 billion (Gabor, 2005).

---

[47] P. C. Stern, T. Dietz, T. Abel, G. A. Guagnano and L. Kalof. 1999. "A Value-Belief-Norm Theory of Support for Social
Movements: The Case of Environmentalism," Human Ecology Review, 6: 2, online at <http://www.humanecologyreview.org/pastissues/her62/62sternetal.pdf>(retrieved November 30, 2006).

[48] Environmental Protection Agency, government of United States of America, "Watershed Approach Framework," online at <http://www.epa.gov/owow/watershed/framework.html> (retrieved November 29,2006).

REFERENCES

BOOKS, ARTICLES AND REPORTS

AAFC. 2005. *Canada's Agricultural Policy Framework. 2005.* Ottawa: Agriculture and Agri-food Canada (AAFC)

Adger, W. N. 2000. Social and Ecological Resilience: are they related? *Progress in Human Geography.* 24(3): 347-364.

Agnew, P. 2005. A Review of Agri-Environmental Policy and Programs in Canada, the United States and the European Union. Masters thesis, University of Guelph, Rural Extension Studies.

Agnew, P. and Filson, G. C. 2004. Landowner Motivations for the Adoption/Non-Adoption of Best Management Practices in the Hobbs-McKenzie Drainage Area and Usborne Township. *Technical Report prepared for the Ausable Bayfield Conservation Authority.* Guelph: University of Guelph.

Amin, S. 1974. *Accumulation of a World Scale: A Critique of the Theory of Underdevelopment.* N. Y.: Monthly Review Press.

Arcury, T. A. and E. H. Christianson. 1990. Environmental worldview in response to environmental problems: Kentucky 1984 and 1988 compared. *Environment Behavior* 22:387–407.

Armitage, D., Bonnet, R. and FitzGibbon, J. 2005. Development of Nutrient Management in Ontario: A review of Progress toward Continuous Improvement. Identifying Strategies to Support Sustainable Agriculture in Canada. Quebec City: Agricultural Institute of Canada.

Babbie, E. 1998. *The Practice of Social Research.* Belmont: Wadsworth Publishing Company.

Baker, S. 2006. *Sustainable Development.* London: Routledge.

Beaulieu, M. 2005. Encouraging the Adoption of Beneficial Management Practices on Canadian Farms: Who Should we Reach out to? Identifying Strategies to Support Sustainable Agriculture in Canada. Quebec: Forum of the Agricultural Institute of Canada.

Belletti, G. Marescotti, A. and Moruzzo, R. Possibilities of the new Italian law on Agriculture, in Durand, G. and Van Huylenbroeck, G. (eds.). 2002. *Multifunctional Agriculture: A New Paradigm for European Agriculture and Rural Development.* Hampshire: Ashgate Pub. 143-168.

Berkes, F., Colding, J. and Folke, C. (eds). 2003. *Navigating Social-Ecological Systems: Building Resilience for Complexity and Change.* Cambridge: Cambridge University Press.

Berkes, F. and Folke, C. 1998. Linking Social and Ecological Systems for Resilience and Sustainability. In Berkes, F. and Folke, C. (eds.) *Linking Social and Ecological Systems: Management Practices and Social Mechanisms* (pp.1-26). Cambridge: Cambridge University Press.

Bernstein, M. H. 2004. *Withouut a Tear: Our Tragic Relationship with Animals.* Urbana: University of Illinois Press.

Black, J. S., Stern, P. C. and Elworth, J. T. 1985.Personal and contextual influences on household energy adaptations. *Journal of Applied Psychology* 70: 3–21.

Blackie, M. and Tuininga, K. 2003. Environment Canada-Ontario Regional Compliance Promotion 2002 Target Sub-watershed Survey Report. Ottawa: Environment Canada.

Bohman, M., Cooper, J., Mullarkey, D., Normile, M. A., Skully, D., Vogel, S. and Young. E. 1999. *The Use and Abuse of Multifunctionality.* Washington: Economic Research Service/USDA.

Bonnett, R., FitzGibbon, J. and Armitage, D. 2005. Development of Nutrient Management in Ontario: Toward Continuous Improvement. Quebec City: Agricultural Institute of Canada.

Bowler, I. R. 1992. *Geography of Agriculture in Developed Market Economies.* N. Y.: John Wiley.and Sons.

Braithwaite, J. and Drahos, P. 2000. *Global business regulation.* New York : Cambridge University Press,

Brethour, C. Mar. 22, 2007. An Economic Evaluation of BMPs for Crop Nutrients in Canadian Agriculture, *Environmental BMP Adoption Workshop.* Mississauga, ON.

Brown, D. G., Johnson, K. M., Loveland, T. R. and Theobald, D. M. 2005. Rural Land Use Trends in the Coterminous United States, 1950-2000, *Ecological Adaptations.* 15(6): 1851-1863.

Bucknell, D., Filson, G. C. and Hilts. S. 2004. Farmer and Non-Farmer Attitudes to Environmental Practices in Agriculture in Two Sub-watersheds of Ontario's Grand River. Guelph: University of Guelph Manuscript.

Bucknell, D. 2002. Rural Quality of Life in a Changing Environment: A Study of the Eramosa-Speed Region. Masters thesis, University of Guelph, Rural Extension Studies.

Buller, H., G. A. Wilson, and Hoell, A. (eds). 2002. *Agri-Environmental Policy in the European Union*. Aldershot: Ashgate Publishing Company.

Buttel, F. H. 2001. Some Reflections on Late Twentieth Century Agrarian Political Economy. *Sociologia Ruralis: Journal of European Society for Rural Sociology.* (48)2: 165-181.

Caldwell, W. 2005. Rural Planning Methods, RPD6280 Notes. Guelph: University of Guelph.

Carson, R. 1962. *Silent Spring*. Cambridge: Houghton Mifflin.

Chekay, D. Nov. 2006. Equivalent Agri-Environmental Farm Planning: An Innovative Tool for Implementing Sustainable Agricultural Practices on a Landscape Scale. *Innovation for Growth: Trends and Successes Redefining Agriculture*. Winnipeg: Agricultural Institute of Canada. P. 39.

Conyers, D. and Hills, P. 1984. *An Introduction to Developing Planning in the Third World*. N. Y.: John Wiley and Sons.

Conca, K. and Debalko, G. (eds.) 2004. Green planet blues : environmental politics 3rd Edition Boulder, Colo. : Westview Press, 2004.

Cooke, S. 2006. *Water Quality in the Grand River: A Summary of Current Conditions (2002-2004) and Long Term Trends*. Cambridge: Grand River Conservation Authority.

Cooke, S. E., and Prepas, E. E. 1998. Stream Phosphorus and Nitrogen Export from Agricultural and Forested Watersheds on the Boreal Plain. *Canadian Journal of Fish and Aquatic Sciences*, 55: 2292-2299.

Corkal, D., Schutzamn, W. C. and Hilliard, C. 2004. Rural Water Safety from the Source to the On-Farm Tap. *Journal of Toxicology and Environmental Health, Part A,* 67(20-22): 1619-1642.

Coward, H. Religious Responsibility. In Coward, H. and Hurka, T. (eds.) 1993. *Ethics and Climate Change: The Greenhouse Effect*. Waterloo: Wilfrid Laurier University Press. Pp. 39-60.

Daneshvary, N., Daneshvary, R. and Schwer, R. K. 1998. Solid-waste recycling behavior and support for curbside textile recycling. *Environment and Behavior* 30:144–161.

Del Mar Delgado, M, Ramos, E., Gallardo, R. and Ramos, F. Multifunctionality and Rural Development: a Necessary Convergence, in Durand, G. and Van Huylenbroeck, G. (eds.). 2002. *Multifunctional Agriculture: A New Paradigm for European Agriculture and Rural Development*. Hampshire: Ashgate Pub. 19-36.

Denhartog, J. Apr. 20, 2007. Farmers provide much more to consumers than food. *The CFFO Commentary.*

Dey, P. Changes in Land Use Patterns and their Implication for the Sustainability of Rural Communities: A Critical Synthesis of Existing Literature. Guelph: Univ. of Guelph PhD in Rural Studies Program.

Durand, G. and Van Huylenbroeck, G. Multifunctionality and Rural Development: A General Framework, in Durand, G. and G. Van Huylenbroeck (eds.). 2002. *Multifunctional Agriculture: A New Paradigm for European Agriculture and Rural Development.* Hampshire: Ashgate Pub. 1-18.

Duff, S., Stonehouse, D. P. , Hilts, S. G. and Blackburn, D. G.. May-June:1991. Soil Conservation Behavior and Attitudes Among Ontario Farmers. *Journal of Soil and Water Conservation,* 46(3): 215-219.

Dunlap, R. E. (1975). The Impact of Political Orientation on Environmental Attitudes and Actions. *Environment and Behaviour*, 7: 428-454.

Environics Research Group. May 2006. *National Survey of Farmers and Ranchers: Ecological Goods and Services..* Prepared for Wildlife Habitat Canada.

Ervin, A. et al. 1982. Factors Affecting the Use of Soil Conservation Practices: Hypotheses, Evidence, and Policy Implications. *Land Economics,* 58(30): 271-292.

Falkenmark, M. 2001. The Greatest Water Problem: the Inability to Link Environmental Security, Water Security and Food Security. *Water Resources Development,* 17(4): 539-554.

Featherstone, M. and Goodwin, B. 1993. Factors Influencing a Farmer's Decision to Invest in Long-term Conservation Improvements. *Journal of Soil and Water Conservation,* 46(5): 365-370.

Filson, G. C., Sethuratnam, S., Adekunle, B. and Lamba, P. 2009. Beneficial Management Practice Adoption in Five Southern Ontario Watersheds, *Journal of Sustainable Agriculture.* 34(2): 229-252.

Filson, G. C. 1983. Class and Ethnic Differences in Canadians' Attitudes Towards Native People's Rights and Immigration, *Canadian Review of Sociology and Anthropology* **20** (4): 454-482.

Filson, G. C. 2004a. Issues and Overview. In Filson, G. C. (ed) *Intensive Agriculture and Sustainability: A Farming Systems analysis* Vancouver: UBC Press. Pp. 3-14.

Filson, G. C. 2004b. Social Implications of Intensive Agriculture. In Filson, G. C. (ed) *Intensive Agriculture and Sustainability: A Farming Systems analysis* Vancouver: UBC Press. Pp. 34-52

Filson, G. C. 2004c. Introduction. In Filson, G. C. (ed.) *Intensive Agriculture and Sustainability: A Farming Systems Analysis* pp. 3-14.

Filson, G. C. 2004d. Environmental Problems associated with Intensive Agriculture. In Filson, G. C. (ed.) *Intensive Agriculture and Sustainability: A Farming Systems Analysis*. Pp, 15-33.

Filson, G. C. 1996. Demographic and Farm Characteristic Differences in Ontario Farmers' Views about Sustainability Policies. *Journal of Agricultural and Environmental Ethics* 9(2): 165-180.

Filson, G. C. 1993. Comparative Differences in Ontario Farmers' Environmental Attitudes. *Journal of Agricultural and Environmental Ethics* 6(2): 165-184.

Filson, G. C. and C. Duke. 2004. Integrating Farming Systems Analysis of Intensive Farming in Southwestern Ontario. In Filson, G.C. (ed.) *Intensive Agriculture and Sustainability: A Farming Systems Analysis* (pp. 177-190). Vancouver: UBC Press.

Filson, G. C. and Friendship, R. M., June 25, 1999. *Final Report, Ontario Pork Project*, Evaluating the Public's Perception of Pork Production. Pork Congress, Stratford, ON.

Filson, G. C. and Sarker, R. Sept. 2008. Assessing the Impacts of the Greenbelt Act in Ontario on Land Values and on Environmental Goods and Services. A Research Proposal. Guelph: University of Guelph.

Filson, G. C., Stonehouse, D. P., Rudra, R. and Voroney, P. 2004. Improving Sustainability and Nutrient Management of an Agricultural Watershed. Final Report, University of Guelph /OMAFRA. Resources Management and the Environment Program.

Filson, G. C., Stonehouse, D. P., Rudra, R., Hilts, S., Caldwell, W. and Duke, C. 2002. Policy Incentives for Soil and Water Conservation. Final Report, University of Guelph/OMAFRA.Resources Management and the Environment Program.

Filson, G. C., Richmond, L. Paine, C. and Taylor, J. 2000. Non-farm Rural Ontario Residents' Perceived Quality of Life. *Social Indicators Research,* 50: 159-186.

Filson, G. C. and McCoy, M.A. 1993. Farmers' Quality of Life: Sorting out the Differences by Class. *The Rural Sociologist,* 13(1): 14-37.

Filson, G. C., Sethuratnam, S. Adekunle, B. and Lamba, P. 2009. Beneficial Management Practice Adoption in Five Southern Ontario Watersheds. *Journal of Sustainable Agriculture.* 33(2): 1-24.

FitzGibbon, J. 2005. *The Environmental Farm Plan: An Approach to the Management of Agricultural Non-Point Source Contamination.*

Identifying Strategies to Support Sustainable Agriculture in Canada. Quebec City: Agricultural Institute of Canada.

FitzGibbon, J., Plummer, R. and Summers, R. 2004. The Ontario Environmental Farm Plan: A Whole Farm Systems Approach to Participatory Environmental Management for Agriculture. In Filson, G. C. (ed.) *Intensive Agriculture and Sustainability: A Farming Systems Analysis* (pp. 162-176). Vancouver: UBC Press.

FitzGibbon, J., Hammel, S. and Matrunec, S. 2002. *Report on Nutrient Management By-Laws in the Province of Ontario.* Report SR9086, Guelph: Ontario Ministry of Agriculture, Food and Rural Affairs.

Frank, B. and Vibrans, A. C. 2003. Uma Visão Integrada da Bacia Hidrográfica. In: B. Frank and A. Pinehiro (eds). *Enchentes na Bacia do Rio Itajai – 20 anos de experiencias* (pp. 191-221). Blumenau: Edifurb.

Fransson, N. and Garling, T. 1999. Environmental Concern: Conceptual Definitions, Measurement Methods, and Research Findings. *Journal of Environmental Psychology. 19:369-382.*

Friedmann, H. 1986. Family Enterprises in Agriculture: Structural Limits and Political Possibilities. In *Agriculture: People and Policies.* Ed. G. Cox, P. Lowe and M. Winter. London: Allen and Unwin, pp. 20-40.

Friedmann, H. and McMichael, P. 1989. Agriculture and the State System: the Rise and Decline of National Agriculture, 1870 to the Present. *Sociologia Ruralis.*29(2): 93-117.

Friedmann, J. 1987. *Planning in the Public Domain: From Knowledge to Action.* Princeton: Princeton University Press.

Funtowicz, S. O. and Ravetz, J. R. 1994. Uncertainty, Complexity and Post-Normal Science. *Environmental Toxicology and Chemistry,* 13(12): 1881-1885.

Gabor, S., Kirby, J. and Avery, A. 2005. Ecological Goods and Services to Support Sustainable Agriculture in Canada. In Agricultural Institute of Canada report, *Identifying Strategies to Support Sustainable Agriculture in Canada* (p. 26). Quebec City: Agricultural Institute of Canada.

Gale J.A., Line, D. E. Osmond, D. L. Coffey, S. W., Spooner, J. and Arnold, J. A. 1993. Evaluation of Experimental Rural Clean Water Program. In Environmental Protection Agency report (section: *EPA-841-R-93-005).* Environmental Protection Agency.

Gale, J. A., D. E. Line, D.L. Osmond, S. W. Coffey, J. Spooner and J. A. Arnold. 1992. *Summary Report: Evaluation of the Experimental Rural Clean Water Program.* National Water Quality Evaluation Project, NSCU

Water Quality Group, Biological and Agricultural Engineering Department. Raleigh, NC: North Carolina State University.

Gallardo, R., Ramos, F., Ramos, E. and del Mar Delgado, M. 2002. *New Opportunities for non-competitive Agriculture,* 169-188.

Gallopin, G. C., Funtowicz, S. O'Connor, M. and Ravetz, J. 2001. Science for the Twenty-First Century: from Social Contract to the Scientific Core. *International Social Science Journal,* 53(168): 219-230.

Gamba, R. J. and Oskamp, S. 1994. Factors influencing community residents' participation in commingled curbside recycling programs. *Environment and Behavior* 26: 587–612.

Garling, T., Fujii, S., Garling, A. and Jakobsson, C. 2003. Moderating effects of social value orientation on determinants of proenvironmental behavior intention. *Journal of Environmental Psychology. 23:1-9.*

Goldie, K. Mar. 22, 2007. Crop Nutrients Council Survey Results—Farmer Attitudes towards BMPs, *Environmental BMP Adoption Workshop.* Mississauga, ON.

Gordova, M. and Davidova, S. 2001. The International Competitiveness of CEEC Agriculture. *The World Economy,* 24(2): 185-200.

Goss, M. J. , Ogilvie, J. R., Filson, G. C., Barry, D. A. and Olmos, S. 2004. Developing Predictive and Summative Indicators to Model Farming Systems Components, In Filson, G. C. (ed.) Intensive Agriculture and Sustainability: A Farming Systems Analysis (pp. 67-80) . Vancouver: UBC Press.

Graham, A. Mar. 22, 2007. Provincial Accomplishments under the Current APF, *Environmental BMP Adoption Workshop.* Mississauga, ON.

Graham, A. Mar. 22, 2007. Ontario Accomplishments in the APF Environment Chapter: *Environmental BMP Adoption Workshop.* Mississauga, ON.

Grand River Conservation Authority. 2002. *A Watershed Plan for the Canagagigue Creek Watershed, Phase I.* Cambridge: Grand River Conservation Authority.

Harvey, D. *Justice, Nature, and the Geography of Difference.* Cambridge, Mass.: Blackwell Publishers, 1996.

Heilig, G. 2003. Multifunctionality of landscapes and ecosystem services with respect to rural development. In *Sustainable Development of Multifunctional Landscapes,* Helming, K. and Wiggering, H. (eds), Springer, Berlin.

Hine, D. W. and Gifford, R. 1991. Fear appeals, individual differences, and environmental concern. *Journal of Environmental Education* 23:36–41.

Hessenhuber, A. 2006. Sustainable Agriculture Implemented by Cross Compliance: A Challenge for the Extension Service. In *Problems and Perspectives of Agricultural Extension Services Development.* Sudak, Crimea, Ukraine. A Conference by Tavria State Agrotechnical Academy, Pennsylvania State University and the National Association of Extension Services of Ukraine.

Hessing, M., Howlett, M., and Summerville, T. 2005. *Canadian Natural Resource and Environmental Policy: Political Economy and Public Policy.* Vancouver: UBC Press.

Hilts, S. 2006, Ontario's Greenbelt: the Farm Perspective. Guelph: *PhD Rural Studies lecture notes.* Guelph: University of Guelph.

Hilts, S. 2006. Farmland Preservation and the Greenbelt. *PhD Rural Studies lecture notes.* Guelph: University of Guelph.

Hoffman, M. and Beaulieu, M. S. 2006. *A Geographical Profile of Manure Production in Canada* Ottawa: Agriculture Division, Statistics Canada

Howell, S. E. and Laska, S. B.. 1992. The changing face of the environmental coalition: A research note. *Environment and Behavior* 24:134–144.

Ice, G. 2004. History of Innovative Best Management Practice Development and its Role in Addressing Ware Quality Limited Waterbodies. *Journal of Environmental Engineering,* 130(6): 684-689.

Isakson, R. 2002. *Payments for Environmental Services in the Catskills: A socio-economic analysis of the Agriculture Strategy.* New York: New York City Department of Environmental Protection.

Jayasinghe-Mudalige, U., Weersink, A., Deaton, R., Beaulieu, M. and Trant. M. 2005. Effect of Urbanization on the Adoption of Environmental Management Systems in Canadian Agriculture. In Statistics Canada, Agricultural Division *Agricultural and Rural Working Paper Series* (No. 73). Ottawa: Statistics Canada, Agricultural Division.

Kaiser, F.G. and Shimoda, T. A. 1999. Responsibility as Predictor of Ecological Behaviour. *Journal of Environmental Psychology. 19:243-253.*

Kirschenmann, F. L. 2008. Food as Relationship. *Journal of Hunger and Environmental* Nutrition. 3(2):; 106-121.

Klubnikin, K. and Causey, D. 2002. Environmental Security: Metaphor for a Millennium. *Journal of Diplomacy and International Relations,* 3(2): 104-133.

Klupfel, E. J. 2000. Achievements and Opportunities in Promoting the Ontario Environmental Farm Plan. *Environments,* 28(1): 21-36.

Lafferty, W. M. and Hovden, E.. 2003. Environmental Policy Integration: Towards an Analytical Framework, *Environmental Politics.* 12(3): 1-22.

Lamba, P. 2006. Factors Affecting the Adoption of Environmental Best Management Practices in Rural Ontario. Masters thesis, Guelph, University of Guelph, Rural Extension Studies.

Lamba, P., Filson, G. C. and Adekunle, B. 2009. Factors affecting the adoption of best management practices in southern Ontario, *The Environmentalist.* 29(1): 64-77.

Lampkin, N. 2002. *Organic Farming* (pp. 579-580). Ipswich: Old Pond Publishing.

Land Directorate. 1985. *Urbanization of Rural Land in Canada: Land Use in Canada, Fact Sheet.* Ottawa: Environment Canada.

Leeuwis, C. 2004. *Communication for Rural Innovation: Rethinking Agricultural Extension, Third Edition* (pp. 64-66). Oxford: Blackwell Publishing.

Legislative Assembly of Ontario, June 3, 2003, Official Report of Debates (Hansard). Standing Committee on General Government, Nutrient Management Act, 2002. G-45.

Leopold, A. 1949. *A Sand County Almanac and Sketches Here and There.* New York: Oxford University Press (reprint, 1989) (page citations are to the reprint edition).

Luzar, E.J. and Diagne, A. 1999. Participation in the Next Generation of Agriculture Conservation Programs: The Role of Environmental Attitudes. *Journal of Socio-Economics,* 28: 335-349.

Mann, S. A. and Dickenson, J. M.. 1978. Obstacles to the Development of a Capitalist Agriculture. *Journal of Peasant Studies.* 5: 466-481.

Marshall, G. 2004. From Words to Deeds: Enforcing Farmers' Conservation Cost-Sharing Commitments. *Journal of Rural Studies,* 20(2): 157-167.

Maruyama, M. 1978. *Endogenous Research and Polyocular Anthropology in Perspectives on Ethnicity.* Ed. R. Holloman and S. Arutiunov, The Hague: Mouton Publisher.

Maruyama, M. 1985. Mindscapes: How to understand specific situations in Multicultural Management. *Asia pacific journal of Management* 2(3)125-149

Maruyama, M. 1995. Individual Epistemological Heterogeneity across cultures and its use in organisations. *Cybernetica,* 37(3):215-249.

Maruyama, M. 1994. *Mindscapes : the epistemology of Magoroh Maruyama* New York : Gordon and Breach.

Maruyama, M. 2004. Polyocular vision or subunderstanding? *Organization Studies*, 25: 467-480.

Marx, K. 1951. *Theories of Surplus Value*. London: Lawerence and Wishart.

Marx, K. 1959. *Capital. Vol. III*. Moscow: Progress Pub.

Marzall, K. 2006. Exploring Environmental Security in Agro-ecosystems: Interactions between Farmers' Mindscape and Landscape. Doctoral Dissertation, University of Guelph, Rural Studies.

Marzall, K. 2003. Environmental Extension: Redefining Human-Environment Interaction Patterns in Agriculture. Report for Qualifying Examination, University of Guelph, Rural Studies.

Mazoyer, M. and Roudart, L. 2006. *A History of World Agriculture: from the Neolithic Age to the Current Crisis. (Histoire des Agricultures du Monde: du neolithique a la Crise Contemporanine.* Paris : Seiul translated by J. H. Membrez) N. Y.: Monthly Review Press.

McCallum, C. 2003. *Identifying Barriers to Participation in Agri-Environmental Programs in Ontario*. Guelph: Christian Farmers Federation of Ontario.

McCuaig, J. D. and Manning, E. W. 1982. *Agricultural Land-Use Change in Canada: Process and Consequences*. Ottawa: Land Directorate, Environmental Canada.

McIntyre, L. Mar. 22, 2007. Ecological Goods and Services, National Survey of Farmers and Ranchers, 2006. Environics Research Group. *Environmental BMP Adoption Workshop*. Mississauga, ON.

McMichael, P. 2009. A Food Regime Genealogy. *Journal of Peasant Studies.* 36(1): 136-169.

McMichael, P. 1992. Tensions between National and International Control of the World Food Order. *Sociological Perspectives. 35(2): 343-365.*

McMichael, P. 2000. *Development and Social Change* 2nd Edition. Thousand Oaks: Pine Forge Press.

Meadows, D. H., Meadows, D. L., Randers, J. and Behrens, W. W., *The Limits to growth : a report for the Club of Rome's project on the predicament of mankind.* New York: Universe Books.

Meilke, K. and Martin, C. (Dec. 12, 2002). The United States WTO Proposal and the Dairy Sector, Ontario Ministry of Agriculture and Food.

Moccia, R. Apr. 15, 2008. OMAFRA/University of Guelph Contract. A Talk by the Associate Vice-President (Research) Agri-Food and Partnerships, Guelph: University of Guelph.

Montpetit, E. 2003. Misplaced Distrust: Policy Networks and the Environment in France, the United States, and Canada. Vancouver: UBC Press.

Montpetit, E. and Coleman. W/ D/. 1999. Policy Communities and Policy Divergence: Agro-environmental Policy in Quebec and Ontario. *Canadian Journal of Political Science.* Vol. 32: 691-714.

Montpetit, E. 2002. Policy Networks, Federal Arrangements, and the Development of Environmental Regulations: A Comparison of Canadian and American Agricultural Sectors. *Journal of Policy, Administration and Institutions.* 15(1): 1-20.

Morris, C. and Potter, C.. 1995. Recruiting the New Conservationists: Farmers' Adoption of Agri-environmental Schemes in the U.K. *Journal of Rural Studies,* 11(1): 51-63.

Morris, J., Mills, J. and Crawford, I. M. 2000. Promoting Farmer Uptake of Agri-Environmental Schemes: the Countryside Stewardship Arable Options Scheme. *Land Use Policy,* 17: 241-251.

Morito, B. 2002. *Thinking Ecologically – Environmental Thought, Values and Policy.* Halifax: Fernwood.

Muller, M., Tagtow, A., Roberts, S. L. and MacDougall, E. 2009. Aligning Food Systems Policies to Advance Public Health. *Journal of Hunger and Environmental Nutrition.* 4(3): 225-240.

Napier, T. and D. Brown. 1993. Factors Affecting Attitudes toward Ground Water Pollution among Ohio Farmers. *Journal of Soil and Water Conservation,* 48(5): 432-439.

NCSU Water Quality Group. 1992. Chapter 3: Perspectives on the Rural Clean Water Program: Survey Results. Raleigh: North Carolina State University.

Neave, P., Neave, E., Wins, T. and Riche, T. 2001. Availability of Wildlife Habitat on Farmland. In McRae, T., Smith, C.A.S. and Gregorich, I. J. (eds.) *Environmental Sustainability of Canadian Agriculture: Report of the Agri-Environmental Indicator Project, a Summary* (pp. 145-56). Ottawa: Research Brank, Policy Branch, Prarie Farm Rehabilitation Administration, Agriculture and Agri-Food Canada.

Nef, J. 1999. *Human Security and Mutual Vulnerability: the Global Political Economy of Development and Underdevelopment.* Ottawa: International Development Research Council.

Nuedoerffer, R. C., Waltner-Toews, D., Kay, J. J., Joshi, D. D., and Tamang, M.S. (2005). A Diagrammatic Approach to Understanding Complex Eco-Social Interactions In Kathmandu, Nepal. *Ecology and* Society. 10(2): 12.

New York City Department of Environmental Protection. 2001. *New York City's 2001 Watershed Protection Program Summary, Assessment and*

Long Term Plan. New York: New York City Department of Environmental Protection.

Newhouse, N. 1990. Implications of attitude and behavior research for environmental conservation. *Journal of Environmental Education* **22**:26–32.

Nilsson, C. and M. Svednmark. 2002. Basic Principles and Ecological Consequences of Changing Water Regimes: Riparian Plan Communities. *Environmental Management,* 30(40): 468-480.

Noe, E., Alroe, H. F. and Langvad, A. S.. 2005. *A semiotic polyocular framework for multidisciplinary research in relation to multifunctional farming and rural development*. Paper presented on the XXI ESRS-Congress in Hungary, August 22-27.

Nord, M., Luloff, A. E. and Bridger, J. C. 1988. The association of forest recreation with environmentalism. *Environment and Behavior.* 30: 235–246.

Nowak, P.J. and Korsching, P.F. 1983. Social and Institutional Factors Affecting the Adoption and Maintenance of Agricultural BMPs. In F. Schaller and G. Bailey (eds) *Agricultural Management and Water Quality* (p. 349-377). Ames: Iowa State University Press.

O'Connor, D. 2002. *Report of the Walkerton Inquiry: A Strategy for Safe Drinking Water*. Toronto: Walkerton Inquiry Commission, Queen's Printer.

O'Connor, M. (ed). 1994. *Is capitalism sustainable? Political economy and the politics of ecology.* New York : Guilford Press

Ollman, B. 1993. *Dialectical investigations.* New York : Routledge.

Oskamp, S., Harrington, M. J., Edwards, T. C., Sherwood, D.L., Okuda, S. M. and Swanson, D. C. 1991. Factors Influencing household recycling behaviour. *Environment and Behaviour,* 23(4): 494-519.

Palys, T. 2003. Research Decisions: Quantitative and Qualitative Perspectives. Third Edition. Scarborough: Thomson Pub.

Pampel, F. Jr. and van Es, J.C. 1977. Environmental Quality and Issues of Adoption Research. *Rural Sociology,* 42(1): 57-71.

Personal Conversation between Paige Agnew and Paul Nairn. Representative of the Ontario Federation of Agriculture. May 5, 2004.

Personal Conversation between Paige Agnew and Tom Prout. General Manager, Ausable Bayfield Conservation Authority. April 28, 2004.

Pfeiffer, W. C. and G. C. Filson. 2004. Future Challenges awaiting the Dairy Industry as the Result of its Management Decision Environment. In

Filson, G. C. (ed.) *Intensive Agriculture and Sustainability: A Farming Systems Analysis.* Vancouver: UBC Press. Pp. 126-144.

Pimentel, D. 2008. *Food, Energy and Society.* Boca Raton: CRC Press.

Pimentel, D., L. Westra and R. F. Noss (eds.). 2000. *Ecological Integrity: Integrating Environment, Conservation and Health.* Washington: Island Press.

Plummer, R., Spiers, A., Summer, R. and FitzGibbon, J. 2007. The Contributions of Stewardship to Managing Agro-Ecosystem Environments. *Journal of Sustainable Agriculture.* 31(3): 55-84.

Poerksen, B. 2003. "At Each and Every Moment, I can Decide Who I am"- Heinz von Foerster on the observed, dialogic life, and a constructivist philosophy of distinction. *Cybernetics and Human Knowing,* 10(3-4): 9-26.

Protecting Ontario's Water, June 27, 2007. *The Globe and Mail.* C01.

Putnam, R. G. 1959. Changes in Rural Land Use Patterns on the Central Lake Ontario Plain. *Canadian Geographer.* 6(2): 60-68.

Rausser, G. C., Bails, K. and Simon, L. K. 2004. Agri-Environmental Programs in the United States and European Union. In Anania, M. E. Bohman, C. A. Carter, and McCalla, A. F. (Eds), *Agricultural Policy Reform and the WTO: Where Are We Heading?* Cheltenham: Edward Elgar. (chap.5- confused over first initials)

Rae, G. and Beale, B. 2008. Thinking Outside the Fence. International Land Stewardship Policy Options for the Canadian Agriculture Sector. CanadaWest Foundation.

Ravetz, J. R. 1993. The Sin of Science—Ignorance of Ignorance. *Knowledge: Creation, Diffusion, Utilization,* 15(2): 157-165.

Redclift, M. 1987. *Sustainable Development: Exploring the Contradictions.* London.: Routledge.

Redman, C. I. 1999. *Human Impact on Ancient Environments.* Tucson: University of Arizona Press.

Richmond, L. A., Filson, G. C. Paine, C.. Pfeiffer, W. C. and Taylor, J. R. 2000. Non-Farm Rural Ontario Residents' Perceived Quality of Life. *Social Indicators Research,* 50 (2): 159-186.

Risse, M. and Tanner, H. S.. 2004-2005. Effects *of Voluntary Agricultural Best Management Practice Implementation on Water Quality: Literature Review.* The University of Georgia: Department of Biological and Agricultural Engineering.

Rostow, W. W. 1962. *The Process of Economic Growth.* N. Y.: Norton.

Rogers, Everett M. 1995. Diffusion of Innovations. Fourth Edition. New York: The Free Press.

Roling, N. 1988. Extension Science: Information System in Agricultural Department. Cambridge: Cambridge University Press.

Rubin, J. 2009. *Why Your World Is About to Get a Whole Lot Smaller.* Toronto: Random House.

Ryan, T. July/August. 2007. Should farmers be compensated for their environmental work? *Grand Actions-The Grand Strategy Newsletter* 12(4): 1-2.

Ryan, T. 2006. Rural Water Quality Program—Sharing the Cost of Clean Water, 6th Annual Grand River Watershed Water Forum: Building the Toolkit for Healthy Waters. Cambridge, ON.

Ryan, T. 2000. The Rural Water Quality Program: the City Pays. In Ogilvie, J. R., Smithers, J. and Wall, E. (eds.). *Sustaining Agriculture in the 21st Century Proceedings of the 4th Biennial Meeting* (pp. 61-67). North American Chapter, International Farming Systems Association. Guelph: University of Guelph.

Ryan T.1999. Best Management Practices and the Bottom Line: Profitability and Adoption of BMPs. Cambridge: Grand River Conservation Authority.

Ryan, T., Email message to Sridharan Sethuratnam, Nov. 17, 2006.

Ryan, T. 1999. The Clean Up Rural Beaches Program: Environmentalism in Action? Masters thesis, University of Guelph, Rural Extension Studies.

Saha, B. 2006. Crop Biotechnology, Structure of Primary Production and Socioeconomic Changes in Rural Communities. Guelph: PhD dissertation in Food, Agricultural and Resource Economics.

Samdahl, D. M. and Robertson, R. 1989. Social determinants of environmental concern: Specification and test of the model. *Environment and Behaviour* **21**: 57–81.

Schakel, J. Transdiciplinarity and Plurality or the Consequences of Multifunctionality for Agricultural Science and Education, in Durand, G. and Van Huylenbroeck, G. (eds.). 2002. *Multifunctional Agriculture: A New Paradigm for European Agriculture and Rural Development.* Hampshire: Ashgate Pub. 225- 324.

Serman, N., and Filson, G. C. 2000. Factors Affecting Farmer's Adoption of Soil and Water Conservation. In *Sustaining Agriculture in the 21st Century* (Proceedings of the 4th Biennial Meeting, North American Chapter, International Farming Systems Association), ed. Ogilvie, J. R., Smithers, J. and Wall, E. 69-78. Guelph: University of Guelph.

Serman, N. 1999. Factors Which Influence the Farmer's Adoption of Soil and Water Conservation Practices of Southwestern Ontario. Masters thesis, University of Guelph, Rural Extension Studies.

Sharov, A. 2001. Pragmatism and Umwelt-Theory. *Semiotica*, 134(1/4): 211-228.

Sly, P.G. 2000. *Charting our Common Future: Science the Ecosystem Approach and Sustainable Development (A Canadian Perspective)*. Ottawa: Rawson Academy of Aquatic Science (p. 49).

Smith, D. and Harland, M. 2005. Agricultural Planning for Environmental Action: a Canadian Perspective on Environmental Farm Planning. In Agricultural Institute of Canada, *Identifying Strategies to Support Sustainable Agriculture in Canada* (pp. 66). Quebec City: Agricultural Institute of Canada. (chap. 1)

Smithers, J. and Furman, M. 2003. Environmental Farm Planning in Ontario: Exploring Participation and the Endurance of Change. *Land Use Policy*, 20(4): 343-356.

Sparling, D. June, 2007. *Five Sectors, Five Futures-Can One Policy Framework Really Work?* Guelph: Institute of Agri-Food Policy Innovation.

Sparling, D. and Laughland, P. 2006. *The Two Faces of Farming*. Guelph: Institute of Agri-Food Policy Innovation.

Stern, P. C. 1992. Psychological dimensions of global environmental change. *Annual Review of Psychology* 43:269–302.

Stern, P. C., Dietz, T., Kalof, L. and Guagnano, G. A. 1995. Values, beliefs and pro-environmental action: Attitude formation toward emergent attitude objects. *Journal of Applied Social Psychology* 25:1611–1636.

Stern, P. C. and Oskamp, S. 1987. Managing Scarce Environmental Resources. In Stokols, D. and Altman, I. (Eds), *Handbook of Environmental Psychology, Vol. 2*. New York: Wiley.

Stonehouse, P., de Vos, G. W. and Weersink, A. 2004. A Whole Farm Systems Approach to Modeling Sustainable Manure Management on Intensive Swine Finishing Farms. In G. C. Filson (ed.) *Intensive Agriculture and Sustainability: A Farming Systems Analysis* (pp. 99-115). Vancouver: UBC Press. (chap.5)

Stonehouse, D. P. 1996. A Targeted Policy Approach to Inducing Improved Rates of Conservation Compliance in Agriculture. *Canadian Journal of Agricultural Economics.* 44(2): 105-120.

Stonehouse, D. And Bohl, M. 1993. Selected Government Policies for Encouraging Soil Conservation on Ontario Cash-grain Farms. *Journal of Soil and Water Conservation,* 48(4): 343-349.

Swagemakers, P. Novelty production: New Directions for the Activities and Role of Farming, in Durand, G. and Van Huylenbroeck, G. (eds.). 2002. *Multifunctional Agriculture: A New Paradigm for European Agriculture and Rural Development.* Hampshire: Ashgate Pub., 189-208.

Traore, N., Landry, R. and Amara, N. 1996. On-farm Adoption of Conservation Practices: The Role of Farm and Farmer Characteristics, Perceptions and Health Hazards. *Land Economics,* 74(1): 114-127.

Tremblay, P., Pronovost, J. and Dumais, M. 2008. *Agriculture and Agri-food: Securing and Building the Future.* Québec: Commission sur l'avenir de l'agriculture et de l'agroalimentaire québécois.

Tucker, M. and. Napier, T. L. 2002. Preferred Sources and Channels of Soil and Water Conservation Information Among Farmers in Three Midwestern US Watersheds. *Agriculture, Ecosystems and Environment,* 92: 297-313.

Van Donkersgoed, E. 2005. The Farmer's Stewardship Plan. *Corner Post #376.*

Van Donkersgoed, E. 2005. Top Issues in Ontario Agriculture. *Corner Post #378.*

Van Donkersgoed, E. 2004. A Wave of Government Intrusion. *Corner Post #341*

Van Heyst, B. 2006. Environmental Characterization of Selected Dead Animal Disposal Methods, Ontario Agriculture and Environment Research Day, Guelph. ON.

Van Liere, K. D. and Dunlap, R. E. 1981. Environmental concern: Does it make a difference how it's measured?. *Environment and Behavior* **13**: 651–676.

Van Liere, K.D. and Dunlap, R.E. 1980. The Social Bases of Environmental Concern: A Review of Hypotheses, Explanations and Empirical Evidence. *Public Opinion Quarterly,* 44: 181-197.

Vergopoulos, K. 1978. Capitalism and Peasant Productivity. *Journal of Peasant Studies.* 5: 446-481.

Wallinga, D. 2009. Today's Food System: How Healthy Is It? *Journal of Hunger and Environmental Nutrition.* 4(3): 251-281.

Wandel, J. 1995. An Analysis of Stability and Change in an Old Order Mennonite Farming System in Waterloo Region, Ontario. Masters thesis, University of Guelph, Geography.

Warnock, J. W. 2003 In Diaz, H. P., Jaffe, J. and Stirling, R. (eds.) *Farm Communities at the Crossroads; Challenges and Resistance.* Regina: Canadian Plains Research Center, University of Regina. Pp. 303-323.

Warnock, J. W. 1987. *The Politics of Hunger: the Global Food System.* Toronto: Methuen.

Warnock, J. W. 1978. *Profit Hungry: the Food Industry in* Canada. Vancouver: New Star Books.

WCED (World Commission on Environment and Development). 1987. *Our Common Future.* Oxford: Oxford University Press.

Weis, T. 2007. *The Global Food Economy: the Battle for the Future of Farming.* Halifax: Fernwood Pub.

Wells, K. 2004. Factors Affecting the Adoption of Best Management Practices in the Canagagigue Creek Sub-Watershed. Masters thesis, University of Guelph, Rural Extension Studies.

White, J., Dalrymple, J. and Hume, D. 2007. *The Livestock Industry in Ontario: 1900-2000, A Century of Achievement.* Brampton: InfoResults Ltd.

Wiggering, H., Muller, K., Werner, A. and Helming, K. 2003. The Concept of Multifunctionality in Sustainable Land Development. In *Sustainable Development of Multifunctional Landscapes*, Helming, K. and Wiggering, H. (eds), Springer, Berlin.

Williams, M. 2003. Livestock Intensification: Community Perceptions of Environmental, Economic and Social Impacts and its Impact on Agricultural Production. Masters thesis, University of Guelph, Rural Planning and Development.

Wilton, B. 2006. *The Role of the 'Working Landscape' in Landscapes facing Multiple Pressures from Society.* Guelph: University of Guelph.

Wilton, B. L. 2005. The Multifunctionality of Rural Landscapes. Report for Qualifying Examination, University of Guelph, Rural Studies, PhD Program.

Wilton, B. Nov. 29, 2005. *The Multifunctionality of Rural Landscapes.* Report for Qualifying Examination, University of Guelph, Rural Studies, PhD Program.

Winson, A. 1992. *The Intimate Commodity: Food and the Development of the Agro-Industrial Complex in Canada.* Toronto: Garamond.

Woodhill, J. and Röling, N. G. The Second Wing of the Eagle: the Human Dimension in Learning our way to more Sustainable Futures. In Röling, N. G. and Wagemakers, M. A. E. (eds.).1998. *Facilitating Sustainable Agriculture: Participatory Learning and Adaptive Management in times of*

Environmental Uncertainty Cambridge: Cambridge University Press. Pp. 46-69.

Wright, E. O. 1978. *Class, Crisis and the State*, London: New Left Books

NEWSPAPER ARTICLES AND FILM

Charlebois, S. and Langenbacher, W/ July 3, 2007. Lush Canadian fields but many Fewer Farmers, *Toronto Star. AA8.*

Filson, G. C. and Furlong, K. April, 1980. "Economic Cycle Signals Boom, Bust" *The Clarion* 4(14).

Kenner, R. *Food, Inc.* 2009. Magnolia Pictures.

Roberts, O. 2005, December 19. No Silver Bells for Small Farms this Holiday. *Guelph Mercury*

Roberts, O. Jan. 2008. Mobilize now or lose ground, farmers told. *Guelph Mercury*

Stevenson, M. 2003, August 2. Bending their Ways. *The Globe and Mail*, section F.

Troubled Waters: Audio Gallery. *CBC This Morning and the Sunday Edition.*

Truscott, A. Dec. 12, 2008. Foreign workers decry 'harsh' dismissals from farms. *Globe and Mail.* A13.

Urquhart, I. (2006, November 29). Rural Rumblings Spell Trouble for McGuinty. *Toronto Star*, A. 21.

Ward, O. Apr. 20, 2008. A Vicious Circle of Misery. *Toronto Star.* A.9.

Webb, M. Oct. 11, 2009. Where they grow our junk food. *Sunday Star.* A1 and A10.

Whittington, L. Sept. 5, 2008. Tories blasted on food safety. *Toronto Star.* A17.

INTERNET SOURCES

Agriculture and Agri-Food Canada. 2004. *Putting Canada First: An Architecture for Agricultural Policy in the 21st Century.* http://www.agr.gc.ca/puttingcanadafirst/pdf/gen_e.pdf Retrieved April 13, 2005, from

Alberta Pork, Federal Cull Breeding Swine Program begins April 14th ,
http://www.cpc-ccp.com/documents/7844CPC   swinepgmlaunch   NRL
FINALApr8.pdf. Retrieved Apr. 19, 2008.

Alternative Land Use Services. 2007. http://www.deltawaterfowl.org/alus
/index.php Retrieved July 26, 2007.

America's Clean Water Foundation. 2005. *OFAER*. http://www.acwf.org
Retrieved April 20, 2005

Benefits Outweigh Costs of National ALUS Program: Study. 2007. Canadian
Federation   of   Agriculture.   http://www.cfa-fca.ca/pages/
index.php?main_id=361. Retrieved July 26, 2007.

Blas, J. Nov. 6, 2008. Another Food Crisis Year Looms, says FAO. Financial
Times. http://www.ft.com/cms/s/0/d125dbf2-ac2c-11dd-aa46-000077b076
58,dwp_uuid=a955630e-3603-11dc-ad42-0000779fd2ac
.html?nclick_check=1 Retrieved Dec. 9, 2008.

Busck-Gravcholt, A. 2001. *Summary of Farmers Landscape Decisions:
Relationships between Farmers' Values and Landscape.* http://
www.fao.org/wairdocs.dead/X6133E.htm Retrieved January 15, 2005.

Canadian Broadcasting Corporation, This Morning and the Sunday Edition.
2005. Troubled Waters. http://archives.cbc.ca/IDD-1-75-1390/science_
technology/great_lakes_pollution/ Retrieved June 3, 2006.

Canadian Environmental Law Association. 2005. *Ontario Water Legislation
FAQs.* Retrieved April 19, 2005, from http://www.ecolawinfo.org
/WATER%20FAQs/ Regulatory%20Context% 20for%20Water/ OntWat
Leg.htm#ontwat_01 *A Vision for the Common Agricultural Policy.*
http://www.defra.gov.uk/farm/capreform/pdf/vision-for-cap.pdf Retrieved
September 3, 2006.

Canadian Federation of Agriculture. 2004. *Impact of Agriculture on the
Economy.,* http://www.cfa-fca.ca/english/agriculture   _in_canada
/agriculture_and_the_economy.html Retrieved October 2, 2004.

Clean Water Act. 2006. http://www.ene.gov.on.ca /en/water/cleanwater
/index.php. Retrieved July 6, 2007.

Clean Water Act. 2006. http://www.ene.gov.on.ca/envision/ water/cwa.htm.
Retrieved Oct. 6, 2006

Clean Water Program. 2007. http://www.cleanwaterprogram.ca /Eligible_
projects.htm. Retrieved July 6, 2007.

Clement, J. Feb. 22, 2008. Québec Report has Challenges for Ontario
http://www.christianfarmers.org/main_news_commentaries/2008comment
aries/Feb_22_Quebec_Report.pdf Accessed Feb. 22, 2008.

Clement, J. Dec. 5, 2006. Is it time to dust off the definition of the family farm? Guelph: *CFFO Commentary.* Christian Farmers Federation of Ontario. http://www.christianfarmers.org/main_news_commentaries/2006comment aries/Dec_5. Accessed Dec. 5, 2006

Clement, J. June 29, 2007. Creating Opportunities: For All Farms, Even the Small Ones. *CFFO Commentary.* Christian Farmers Federation of Ontario. http://www.christianfarmers.org/main_news_commentaries/2008comment aries/June_29. Accessed June 29, 2007.

Cross Compliance. 2006. http://www.crosscompliance. org.uk Retrieved September 3, 2006,

Cultural Landscape. (European Union) http://www.pcl-eu.de/project/landscape /index.php. Accessed Nov. 21, 2006.

Dairy Quality Assurance Program. 2005. www.dqacneter.org/fivestar.htm Retrieved April 20, 2005,

De Leeuw. J. 2004. *Netherlands Agri-environmental Policy.* www.fao.or.wairdocs.lead/X6133E/X6133Eoo.htm Retrieved January 6, 2005,

Delta Waterfowl. http://www.deltawaterfowl.org/pr/2007/070416_ALUS.php. Retrieved Mar. 30, 2008.

Environment Canada. 2003. *Beaver River Water Quality Improvement Project.* 2005 from http://www.ec.gc.ca/ecoaction/success_display_ stories_e.cfm ?story_ID=1203018. Retrieved September 5.

Environmental Farm Planning. 2005. http://www.agr.gc.ca/progser/ ps_efppef_e.phtml Retrieved November 22, 2005.

Environmental Sustainability of Canadian Agriculture: Report of the Agri-Environmental Indicator Project. 2001. http://www.agr.gc.ca/env /naharp-pnarsa/index_e.php?page=aei Retrieved March 13, 2006.

Erosion Technology Group, Nov. 2008. *Who Controls Nature? Corporate Power and the Final Frontier in the Commodification of Life.* http://www.etcgroup.org. Retrieved Dec. 9, 2008.

European Union. 2004. *Overview of the European Union Activities-Agriculture.* http://europa.eu.int/pol/agr/overview_en.htm Retrieved April 16, 2005.

Farmasyst. 2005. *Farm*A*Syst Home*A*Syst.* http://www.uwex.edu/ farmasyst Retrieved April 18, 2005.

Goddard, E. and Unterschultz, J. 2004. BSE leaves Canada's beef industry mired in uncertainty *Express News.* http://www.expressnews. ualberta.ca/ article.cfm?id=5514. Retrieved Oct. 7, 2009.

Grand River Conservation Authority. 2004. *Rural Water Quality Program.* http://www.grandriver.ca/index/document.cfm?Sec=25andSub1 =0andsub2=0 Retrieved April 20, 2005.

Grand River Conservation Authority. 2006. *ARTICLE.* http://www.grandriver.ca/index/document.cfm?Sec=25andSub1=0andsub2=0 Retrieved May 31, 2006.

Grand River Conservation Authority. 2004. *Watershed Report 2004-10-24.* http://www.grandriver.ca/WatershedReportCard/ 2004_Fall_Grand.pdf Retrieved October 24, 2004.

Génier, N.A. 2005. *Raisin Region Conservation Authority (RRCA).* http://www.rrca.on.ca/contact/CONTACT.HTM. Retrieved January 19, 2005.

Guelph Water Management Group: http://www.uoguelph.ca/gwmg/ wcp_ home/Pages/M_ws.htm Retrieved Jan. 11, 2007.

Higgins, E. 1998. *Whole Farm Planning: A Survey of North American Experiments.* http://www.winrock.org/wallacecenter/ documents/pspr09. pdf Retrieved: November 3, 2004.

Lake Simcoe Region Conservation Authority. 2005. *State of the Watershed Report: Black River Sub-watershed.* http://www.lsrca.on.ca /PDFs/bk1b. pdf#search='black%20river%20ministry%20of%20the%20environment'. Retrieved September 3, 2005.

Maitland Valley Conservation Authority. 2005. Maitland River Conservation Authority. http://www.mvca.on.ca/. Retrieved May 6, 2005.

McGee, B. 2005. *Ontario Gross Domestic Product (GDP) for Selected Industries, 2004* ($ million). http://www.omafra.gov.on.ca/english/stats/ food/gdp.html Retrieved Dec. 5, 2005.

McGuinty Government speaks up for Ontario Farmers. Agriculture Minister Presents Province's Position at National and International Meetings. 2006. http://www.omafra.gov.on.ca/english/infores/releases/ 2006/ 062306.html Retrieved August 1, 2006.

Miller, J. C. and Coble, K. H. 2005. Cheap Food Policy: Fact or Rhetoric? Agricultural Economics Department, Mississippi State University. http://130.18.41.19/research/Cheap-Food-Policy-Paper.pdf Retrieved July 6, 2007.

Ministry of the Environment. 2004. *Safe Drinking Water Act, 2002.* http://www.ene.gov.on.ca/envision/water/sdwa Retrieved April 13, 2005.

Ministry of Environment. 2003. *Protecting Ontario's Drinking Water: Toward a Watershed Based Source Protection Planning Framework, Final Report, 2003.* http://www.ene.gov.on.ca/envision/techdocs/ 383e.pdf Retrieved April 22, 2006.

Ministry of Municipal Affairs and Housing. 2004. *Greenbelt Plan 2005.* http://www.mah.gov.on.ca/userfiles/HTML/nts_1_22087_1. html#greenbelt Retrieved April 13, 2005.

National Agriculture Centre. 2005. http://www.countrylife.oirg.uk/leaf (chap.3) Retrieved April 18, 2005.

National Farm Stewardship Program. 2004. Retrieved, Nov. 25, 2005, from http://www.agr.gc.ca/progser/ps_nfsp_e.phtml (chap. 1)

National Farm Stewardship Program, 2004. http://www.bcac.bc.ca /documents/NFSP_Terms_and_Conditions_final.pdf Retrieved Nov. 7, 2006.

National Farmers Union Region 3. 2003. *Report on Farm Policy Issues to the Ontario Liberal Caucus, Toronto, Ontario.* http://www.nfu.ca/on/briefs/ Ontario_Liberal_Caucuas_April.brief.pdf Retrieved May 18, 2006.

Ontario Cosmetic Pesticides Ban. 2009. (http://www.ene.gov. on.ca/en /land/pesticides/factsheets/fs-agriculture.pdf), retrieved Oct. 14, 2009.

Ontario Environmental Registry. 2003. *The Purpose of the Environmental Bill of Rights.*, http://www.ene.gov.on.ca/envision/env_reg/ebr/ english/ebr_ info/purpose.htm Retrieved April 13, 2005.

Ontario Farm Environmental Coalition. 2006. http://www.ofa. on.ca/site/ main.asp?pic=../cutting/maintit_issues.jpgandline= 900andInc=../policyissues/issues/Source%20Water%20Protection.htm Retrieved May 31, 2006.

Ontario Farm Environmental Coalition. 2006. Overhwelming Uptake of Environmental Farm Plans by Ontario's Farmers. http://www.vhqfoods.ca /consumer-issues/environmental-farm-plans.aspx. Retrieved Dec. 16, 2008.

Ontario Federation of Agriculture (OFA). 2005. *New EFP Launched in Ontario.* http://www.ofa.on.ca/site/main.asp?pic=../cutting/ maintit_news. jpgandline=1000andInc=../newsevents/newsArticles/2005/april/New%20E FP%20program%20launched%20in%20Ontario.htm Retrieved April 19, 2005.

Ontario's Landowners Association, 2008. Rural Revolution: This Land is Ours, OntarioBack Off Government. http://ruralrevolution.com/website/ Retrieved Dec.11, 2008.

Ontario Ministry of Agriculture and Food. 2005. *Healthy Futures for Ontario Agriculture.* http://www.gov.on.ca/OMAFRA/ english/hfoa Retrieved April 14, 2005.

Ontario Ministry of Agriculture and Food. 2005. *Healthy Futures for Ontario Agriculture: Questions and Answers.* http://www.omafra.gov.on.ca/scripts /english/hfoa/qanda.asp Retrieved April 19, 2005

Ontario Ministry of Agriculture and Food. 2005. *Nutrient Management.* http://www.gov.on.ca/OMAFRA/english/agops/index.html   Retrieved April 13, 2005.

Ontario Ministry of Agriculture and Food. 2005. *Water Management.* http://www.gov.on.ca/OMAFRA/english/environment/water/ legislation.htm. Retrieved April 13, 2005.

Ontario Ministry of Agriculture and Food. 2003. *Best Management Practices.* http://www.gov.on.ca/OMAFR/english/   environment/bmp/series.htm Retrieved April 20, 2005..

Ontario Ministry of Agriculture and Food. 1998. *Ontario Environmental Farm Plan.* http://www.gov.on.ca/OMAFRA/english/ environment/efp/efp.htm Retrieved April 20, 2005.

Ontario Ministry of Agriculture Food and Rural Affairs and Agriculture and Agri-Food Canada. 2002. *Agriculture-Environmental Programs in Ontario.* http://res2.agr.ca/initiatives/manurenet/env_prog/ag_env.html Retrieved April 20, 2005.

Ontario Ministry of Agriculture Food and Rural Affairs and Agriculture and Agri-Food. 2002. *Environmental Sustainability Initiative: Background and Objectives.* http://res2.agr.ca/initiatives/manurenet/ env_prog/esi/ esimenu.html Retrieved April 14, 2005.

Ontario Ministry of Agriculture Food and Rural Affairs and Agriculture and Agri-Food. 2002. *Great Lakes Water Quality Program: 1989-1994 Overview.* http://res2.agr.ca/initiatives/manurenet/env_prog/glwq/glwql.html #Remdedial%20Actions Retrieved April 14, 2005.

Ontario Ministry of Agriculture Food and Rural Affairs and Agriculture and Agri-Food Canada. 2002. *Land Management Assistance Program: Background.* http://res2.agr.ca/initiatives/manurenet/env_prog/lmap/ lma pback.html Retrieved April 14, 2005.

Ontario Ministry of Agriculture Food and Rural Affairs and Agriculture and Agri-Food. 2002. *National Soil Conservation Program.* http://res2.agr.ca/ initiatives/manurenet/env_prog/nscp/nscpmenu.html Retrieved April 14, 2005..

Ontario Ministry of Agriculture Food and Rural Affairs and Agriculture and
Agri-Food Canada. 2002. *Report of the Agreement Management
Committee on the Tentative Program Area Priorities for Green Plan Agri-
Food Funding in Ontario*. http://res2.agr.ca/initiatives/ manurenet/
env_prog/gp/gpres/ gprog92a.html Retrieved April 20, 2005.

Ontario Ministry of Agriculture Food and Rural Affairs and Agriculture and
Agri-Food, 2002. *Soil and Water Environmental Enhancement Program*.
http://res2agr.ca/initiatives/manurenet/env_prog/sweep/sweephom.html
Retrieved April 14, 2005.

Ontario Ministry of Agriculture Food and Rural Affairs and Agriculture and
Agri-Food, 2002. *Soil and Water Environmental Enhancement Program,
Executive Summary*. http://res2.agr.ca/initiatives/manurenet/env_ prog
/sweep/rep5.html#Executive%20Summary Retrieved April 14, 2005.

Ontario Pork, Dec. 1, 1998. *Ontario Pork Producers Rally Together at Queens
Park*. http://www.agpub.on.ca/pc/pr/op1201.htm Retrieved Aug. 19, 2008.

Ontario Soil and Crop Improvement Association, Environmental Programs
Update Report, April 18, 2005-Mar. 31, 2007. http://www.ontarios
oilcrop.org/User/Docs/Programs/March_31_2007_Environmental_Progra
ms_Update_Report.pdf Retrieved Apr. 24, 2007.

Ontario Soil and Crop Improvement Association, 2006. *Programs*.
http://www.ontariosoilcrop.org/cms/en/Programs.aspx.   Retrieved
November 16, 2006.

Ontario Soil and Crop Improvement Association. 2004. *EFP dollars Fully
Committed*. http://www.ontariosoilcrop.org/EFP.htm Retrieved April 25,
2004, from

Pollution Inspection and Investigation. http://www.qc.ec.gc.ca/dpe/ Anglais
dpe_main_en.asp?insp_lcpe_main Retrieved Mar. 2, 2008.

Places to Grow.. http://www.pir.gov.on.ca/userfiles/ HTML/cma_4_40890_
1.html Retrieved Apr. 26. 2006.

Pomeroy, R. Apr. 11, 2008. Food riots to worsen without global action: U.N.
Thomson Reuters.http://ca.reuters.com /article/topNews/ idCAL11907845
20080411 Retrieved Apr. 18, 2008.

Preibisch, K. Nov. 25, 2008. Victory for farm workers in Ontario. *Guelph
Mercury. http://news.guelphmercury.com/Opinions/article/408476* Retri
eved Nov. 25, 2008.

Programs in Ontario. Retrieved April 20, 2005, from http://res2.agr.ca/
initiatives/manurenet/env_prog/ag_env.html.

Protecting Our Drinking Water Sources. http://www.sourcewater.ca/
Retrieved Jun 15, 2007

Reinhart,A., Oct. 10, 2009. Marsh Madness, *Globe and Mail.* M1 and 4.

Risse, L. M. and Cabrera, M. L. 2002. *Land Application of Manure for Beneficial Reuse'* National Animal Waste Management Centre. http://www.cals.ncsu.edu/waster_mgmt/natlcentre/papers.htm Retrieved July 27, 2004.

Rose, Jeff. Feb. 1995. Who do you want in your face? Me or the Government? A lecture given to Agri-food Communities, Guelph: Ontario Agriculture College.

Ryan, T. E. 1999. *The Clean Up Rural Beaches: Environmentalism in Action?* http://www.collectionscanada.ca/obj/s4/f2/ dsk2/ftp03/ MQ40437.pdf Retrieved April 14, 2005.

Saugeen Conservation. 2008. Conservation Authorities Act. http:// www.svca.on.ca/caa.htm Retrieved Dec. 17, 2008.

Statistical Package for the Social Sciences (SPSS). 2005. *SPSS Lesson 3: A Few Basic Statistics.* Retrieved October 18, 2005. frohttp://www.sfu.ca/~palys/SPSS%20Lesson%203.pdf#search='chi%20s quare%20cross%20tabulations'.

Statistics Canada. 2001. *Census of Agriculture.* http://www.statcan.ca /english/agcensus2001/index.htm; Retrieved October 6, 2004.

Statistics Canada. 2008-10-31. Total farm area, land tenure and land in crops, by province (Census of Agriculture, 1986 to 2006) (Ontario). http://www40.statcan.gc.ca/l01/cst01/agrc25g-eng.htm. Retrived Dec. 9, 2008.

Statistics Canada. 2001. *Population Statistics.* http://www12.statcan.ca /english/profil01/Details Retrieved June 10, 2005.

Statistics Canada. 2001. *Population Structure and Change in Predominantly Rural Regions, Rural and Small Town Canada Analysis Bulletin* 2(2). http://www.statcan.ca/english/freepub/21-006-XIE/free.htm Retrieved: April 12, 2002.

Statistics Canada. 2002. *Sharp Decline in Number of Farms in Ontario.* Retrieved June 4, 2002, from http://www.statcan.ca /english/agcensus2001 first/regions/farmon. htm#7

Statistics Canada. 2001. *Urban Consumption of Agricultural Land, Rural and Small Town Canada Analysis Bulletin,* 3(2). http://www.statcan.ca english/freepub/21-006-XIE/free.htm Retrieved April 12, 2002.

Stevens, N. Jan. 12, 2007. Paying for Preserving the Environment and Protecting Habitat. *The CFFO Commentary.* www.christianfarmers. org.html. Retrieved Jan. 12, 2007.

Stevens, N. June 22, 2007. Tackling the Concept of Need in Safety Net Design. C*FFO Commentary,* www.christianfarmers.org.html. Retrieved June 22, 2007.

Stevens, N. Jan. 18, 2008. Dealing with Change through Leadership, *CFFO Commentary.* www.christianfarmers.org.html. Retrieved Jan. 18, 2008.

Tietz, J. 2006. Boss Hog. http://www.rollingstone.com/ politics/story/ 12840743/porks _dirty_secret_the_ nationals_top _hog_producer_is _also _one_of_americas_worst_polluters Retrieved Aug. 19, 2008.

TexaSoft. 2004. *Mann-Whitney Test.* http://www.texasoft. Com/winkm ann.html. Retrieved November 3, 2004.

U.S. Environmental Protection Agency. 2003. *Polluted Runoff (Non-point Source Pollution).* http://www.epa.gov/OWOW/NPS/ qa.html. Retrieved April 13, 2005.

Vellema, S. 2001. Institutional Modalities in Contract Farming: the case of fresh asparagus in the Philippines. Retrieved Jan. 21, 2008.

Vucheva, E. 26.09.2008. 'Laissez-faire' capitalism is finished, says France, EU Observer, http://euobserver.com/9/26814. Retrieved Oct. 10, 2008.

Warnock, J. W. 2002. *Food and Agriculture.* http://www.johnwarnock.ca/ foodandagriculture.html Retrieved Dec. 20, 2007.

Water Quality. http://www.ene.gov.on.ca/envision/techdocs/ 4383e.pdf Retrieved April 22, 2006.

Water Management. 2004. http://www.cals.ncsu.edu/ waster_mgmt/natlcentre /papers.htm; Retrieved July 27, 2004.

What is Multifunctional Agriculture? 2008. http://www.maff.go.jp/soshiki/ kambou/joutai/onepoint/public/ta_me.html. Accessed November 7, 2008.

Wikipedia (2008) Common Agricultural Policy, http://en.wikipedia.org/wiki/ Common_Agricultural_Polcy. Accessed, January 5, 2008.

Wilson, K. 2001. Census of Agriculture and Policy Analysis Branch, OMAFRA. http://www.omafra. gov.on.ca/english/stats/ county /index.html. Retrieved august 15, 2005.

Worrell, R. and Appleby, M. C. 2000. Stewardship of Natural Resources: Definition, ethical and practical aspects. *Journal of Agricultural and Environmental Ethics.* 12: 263-277.

Yosemite Valley Plan. 2004. *Definitions of Best Management Practices and Stewardship.* http://www.nps.gov/yose/planning Retrieved May 23, 2005.

INDEX

Q

quality assurance, 37, 53
quality of life, 15, 23, 30, 39, 42, 57, 64, 68, 71, 73, 76, 79, 81, 84, 85, 87, 104, 106, 109, 119, 130, 132, 137, 141, 146
quality production, 28
quantitative technique, 87
quotas, 49

R

range, 16, 26, 42, 51, 69, 95, 157, 164, 170
ratings, 29, 156
rationality, 55
raw materials, 8
reactions, 157, 166
reality, 26, 122, 149
reason, 13, 54, 58, 67, 121, 122, 125, 128, 137, 150
recession, 1, 8, 40, 145, 146
reciprocity, 62
recognition, 16, 20, 52, 150
recommendations, iv, 32, 41, 143, 163
recreation, 13, 27, 42, 121, 148, 164, 184
recycling, 175, 179, 184
reflection, 21, 160
region, 24, 25, 29, 57, 65, 66, 67, 68, 69, 71, 72, 73, 74, 77, 123, 146, 149, 165
Registry, 157, 159, 166, 194
regression, 125, 126, 127, 128, 132, 149
regression equation, 132
regression model, 126, 149
regulation, 31, 32, 33, 38, 41, 42, 48, 50, 54, 55, 58, 59, 73, 75, 77, 78, 83, 88, 100, 101, 102, 103, 108, 115, 125, 131, 137, 141, 145, 153, 167, 174
regulations, 31, 32, 37, 41, 50, 51, 55, 57, 60, 70, 71, 78, 81, 83, 86, 100, 101, 102, 103, 108, 123, 124, 125, 126, 131, 132, 147, 148, 150, 151, 153, 158
regulatory oversight, 167
Rehabilitation Act, 46

relationship, 6, 10, 13, 26, 60, 84, 99, 105, 106, 108, 109, 115, 116, 117, 123, 125, 128, 132, 137, 141, 148, 158
reliability, 105, 123
religion, 67, 77, 79, 80
remediation, 84, 160, 162
remote sensing, 84, 170
renewable energy, 154, 160
representativeness, 12
re-regulating, 32
resentment, 118, 153
reserve currency, 4
resilience, 3, 14, 22, 23, 26, 34
resistance, 144
resource management, 31
resources, 5, 6, 30, 31, 42, 45, 50, 60, 61, 76, 84, 108, 122, 153, 155, 161, 170
respect, 5, 14, 60, 64, 66, 78, 79, 121, 145, 150, 151, 179
restructuring, 7
retirement, 48, 55, 61, 80, 106, 110
returns, 43, 46
revenue, 11, 100, 141
risk, 16, 32, 39, 41, 44, 46, 62, 79, 84, 144, 150, 162, 167, 169, 170
risk management, 16, 32, 44, 167
runoff, 32, 45, 110, 124, 166
rural areas, 5, 29, 109, 154
rural development, 28, 48, 179, 184
rural people, 8, 14, 57, 64, 79, 87, 146, 148, 151, 152
rural population, 69

S

safety, 16, 23, 44, 163
sales, 15, 59, 77, 80, 95, 105, 108, 109, 110, 114, 115, 123, 124, 125, 126, 128, 130, 133, 134, 141, 149, 150, 153
sampling, 63, 69, 104, 105, 127
sampling error, 69, 104, 105, 127
school, 64, 70, 96, 153
search, 5, 193, 197
security, 2, 5, 8, 12, 13, 14, 17, 18, 19, 20, 22, 23, 24, 25, 26, 31, 33, 48, 104, 143, 154, 155, 156, 161, 170

T